LIFE SAVING DRUGS

The Elusive Magic Bullet

RSC Paperbacks

RSC Paperbacks are a series of inexpensive texts suitable for teachers and students and give a clear, readable introduction to selected topics in chemistry. They should also appeal to the general chemist. For further information on all available titles contact:

Sales and Customer Care Department, Royal Society of Chemistry,
Thomas Graham House, Science Park, Milton Road, Cambridge CB4 0WF, UK
Telephone: +44 (0)1223 432360; Fax: +44 (0)1223 426017; E-mail: sales@rsc.org

Recent Titles Available

The Science of Chocolate
By Stephen T. Beckett
The Science of Sugar Confectionery
By W.P. Edwards
Colour Chemistry
By R.M. Christie
Beer: Quality, Safety and Nutritional Aspects
By P.S. Hughes and E. D. Baxter
Understanding Batteries
By Ronald M. Dell and David A.J. Rand
Principles of Thermal Analysis and Calorimetry
Edited by P.J. Haines
Food: The Chemistry of Its Components (Fourth Edition)
By Tom P. Coultate
Green Chemistry: An Introductory Test
By Mike Lancaster
The Misuse of Drugs Act: A Guide for Forensic Scientists
By L.A. King
Chemical Formulation: An Overview of Surfactant-based Chemical Preparations in Everyday Life
By A.E. Hargreaves
Life, Death and Nitric Oxide
By Antony Butler and Rosslyn Nicholson
A History of Beer and Brewing
By Ian S. Hornsey
The Science of Ice Cream
By C. Clarke

Future titles may be obtained immediately on publication by placing a standing order for RSC Paperbacks. Information on this is available from the address above.

RSC Paperbacks

LIFE SAVING DRUGS
The Elusive Magic Bullet

JOHN MANN
Queen's University Belfast, UK

ISBN 0-85404-634-8

A catalogue record for this book is available from the British Library

Published by The Royal Society of Chemistry,
Thomas Graham House, Science Park, Milton Road,
Cambridge CB4 0WF, UK

Registered Charity Number 207890

For further information see our web site at www.rsc.org

Typeset by Macmillan India Ltd., London, UK
Printed by TJ International, Padstow, Cornwall, UK

Preface

At the start of the 20th century, there were only a handful of effective drugs: morphine, quinine, cocaine, aspirin, and a few inorganic salts for gastrointestinal problems. A serious bacterial infection and most cancers would usually be a death sentence for the patient. One hundred years later, there are literally thousands of efficacious drugs – bacterial and viral infections (including HIV) can be successfully treated, and even certain cancers can now be cured. This book describes the evolution of these life-saving drugs that have so revolutionised the treatment of disease, and emphasises the roles played by those who discovered these molecules.

The three main chapters deal with drugs for the treatment of bacterial and viral infections, and cancer; then, the final chapter reveals the new advances that have been facilitated by our growing understanding of the genetic basis of disease. It was never my intention to write a research text, although the bibliography will allow readers to gain access to key research papers; the book is also not intended to compete with the standard textbooks of medicinal chemistry. It should, however, provide an excellent source of background material for students of medicinal chemistry, pharmacy, and even medicine and pharmacology.

The first edition of the book was written primarily for the popular science market; hence, it could not contain chemical structures. It was well-received, with excellent reviews in both *Nature* and the *Times Literary Supplement*; but without the chemical structures, it was impossible to give due prominence to all the fabulous molecular species that have so improved our quality of life and chances of survival. This new edition has been completely updated and expanded, and includes the chemical structures and modes of action for most of the drugs discussed. I hope that it will be of interest to anyone who wants to know more about the molecular entities that comprise the life-saving drugs.

It is a pleasure to thank Liz Hunter and Ian Gibson for the illustrations, and the staff at the RSC for their help during the production of the book.

for Cressida and Octavia

Contents

Chapter 1

The Elusive Magic Bullet: Introduction

Theophrastus Philippus Aureolus Bombastus von Hohenheim would not have been surprised that arsenic was a major component of the first effective treatment for syphilis. This 16th-century alchemist and physician, better known as Paracelsus, is reputed to have acquired his skills from barber surgeons, alchemists and gypsies, although he acquired notoriety after his appointment as Professor of Medicine in Basel. He was renowned for his use of mercury, arsenic, antimony and tin salts for the treatment of syphilis, intestinal worms and sundry other conditions endemic in mediaeval Europe – although it is probable that he killed more patients than he cured with these very toxic metal salts. Four hundred years later, Paul Ehrlich discovered the arsenic-containing drug '606' (later called Salvarsan), the first true antisyphilitic and immediately hailed as a 'magic bullet.' It was to remain the mainstay of treatment until the arrival of another 'wonder drug' of the 20th century – penicillin.

Paracelsus rejected the druglore (mainly plant-based) laid down by the famous Greek physicians Dioscorides and Galen, and believed instead in the so-called 'Doctrine of Signatures.' The main belief was that the shape of a plant revealed the disease or organ that would benefit from therapy involving extracts of that particular plant. Thus, the tooth-shaped seeds of henbane were given for toothache, brain-shaped walnuts for headache and liverwort for liver complaints. His prescribing also owed much to the Arab preference for sulfur, mercury and common salt – the essence of combustibility, fluidity and earthiness – coupled with a firm belief in magic, necromancy and astrology.

These strange ideas (at least in contemporary terms) should not, however, overshadow his major contribution to medicine, which was his forthright belief that each disease had a specific cause and its own remedy. There is an interesting congruence between this belief and the later triumphs of Paul

Ehrlich, who invented just such a treatment for syphilis caused by the organism *Treponema pallidum.*

Paul Ehrlich was born in March 1854 and was brought up in Upper Silesia, in the town of Strehlen. The son of an innkeeper, he received a major part of his education at the University of Breslau. Here, he developed a fascination for the properties of aniline dyes, then a very important product of the German fine chemicals industry. Even in early experiments, he was able to show how certain dyes could help in the identification of cells and for the definition of their fine structure. His secretary of many years, Martha Marquardt, records in her excellent biography of Ehrlich that he was visited in his laboratory one day by Robert Koch, who had become famous through his studies on the causative organisms of anthrax, tuberculosis, diphtheria and several other common diseases. Ehrlich's teacher apparently introduced him to this honoured guest with the words: "That is 'little Ehrlich'. He is very good at staining, but he will never pass his examinations."

Ehrlich did pass his exams and then completed his studies at the University of Leipzig, graduating in 1878 (aged 24) with a doctorate in Medicine. His thesis was, perhaps not surprisingly, entitled: 'Contributions to the Theory and Practice of Histological Staining. Part I. The Chemical Conception of Staining. Part II. The Aniline Dyes from Chemical, Technological and Histological Aspects'. In it, he emphasised the value of an understanding of the chemistry involved in staining organisms, in particular, the specificity of these processes. This search for specific binding (and killing) agents for microorganisms was the central theme for the rest of his research career.

But before research once again dominated his life, he had to make a living and for several years he worked as a house physician in various hospitals. By all accounts, he was a kind and effective clinician, although his research on dyes was never far from his mind. He developed a diagnostic test for the acute phase of typhoid fever, but his major triumph in 1882 was a specific stain for the tubercle bacillus – recently identified by Robert Koch as the cause of tuberculosis – thus facilitating detection of the organism in sputum (saliva) samples. This research undoubtedly led to his own infection by the organism, and in 1887, he began to exhibit the classic symptoms of pulmonary tuberculosis. His illness was not life-threatening but necessitated a two-year sojourn in Egypt, away from the stresses and harsh winters of Germany.

He returned in 1889, completely recovered, and accepted an appointment at the Institute for Infectious Diseases in Berlin directed by Robert Koch. Ehrlich continued to develop his ideas about the specificity of chemicals for particular cells and for their structural components, but he also became heavily involved in the development of methods for the accurate measurement of the potency of diphtheria antitoxin. This seminal work is often overlooked owing to the subsequent excitement over Salvarsan, but it can be viewed as a key part of the evolution of his ideas about chemotherapy.

The bacteriologist Emil von Behring had discovered specific antitoxins in the blood serum of animals infected with sub-lethal doses of diphtheria bacilli, but he was unable to produce samples with reproducible levels of activity. Ehrlich not only showed how to raise high concentrations of antitoxins in horses by successive injections of diphtheria bacilli but also developed the methods necessary for the measurement of the potency of the individual serum samples. The treatment of patients then became both safe and effective, and the fact that his methods are still in use today provides a cogent testimony to his skill and ingenuity.

These studies reinforced his belief in what he termed the *side-chain theory* of interactions. He believed that within each cell, there were large amounts of 'protoplasm' with projections (the 'side-chains'), whose normal physiological function was to interact with the essential chemicals involved in various life processes. These same side-chains could also have a specific affinity for toxins, and once these had been attached, the side-chain lost its capacity to function normally. To compensate, the cell now produced extra copies of the whole side-chain, which were released into the bloodstream and could act as antitoxins. He initially called the combining moiety on the side-chain a *haptophore* and the toxin was the *toxophore*, but later used the expression *receptor* to describe the site at which a foreign organism or drug interacted with the cell. This term is now central to pharmacological terminology. His views on what we would now call *immunotherapy* and *chemotherapy* are probably best summarised in his own words:

"What makes serum therapy so extraordinarily active is the fact that the protecting substances of the body are products of the organism themself, and that they act purely *parasitotropically* and not *organotropically* (i.e., against the body). Here we may speak of *magic bullets* which aim exclusively at the dangerous intruding parasites, strangers to the organism, but do not touch the organism itself and its cells... . But we know of a number of infectious diseases...where serum therapy either does not work at all... . I call attention especially to malaria, to the diseases caused by trypanosomes... . In these cases chemical substances must come to aid the treatment. Instead of serum therapy, *chemotherapy* must be used."

This work in the newly emerging area of immunology occupied Ehrlich for more than a decade, and as a consequence, his research on drug design did not begin in earnest until the turn of the century. At that time, the only drug that had curative properties against a disease (malaria) was quinine and this had been in use since the 17th century, generally in the form of Jesuits' powder, an extract of the bark of the cinchona tree. Ehrlich had been involved with one study (in 1891) of the effect of the dye methylene blue on malaria,

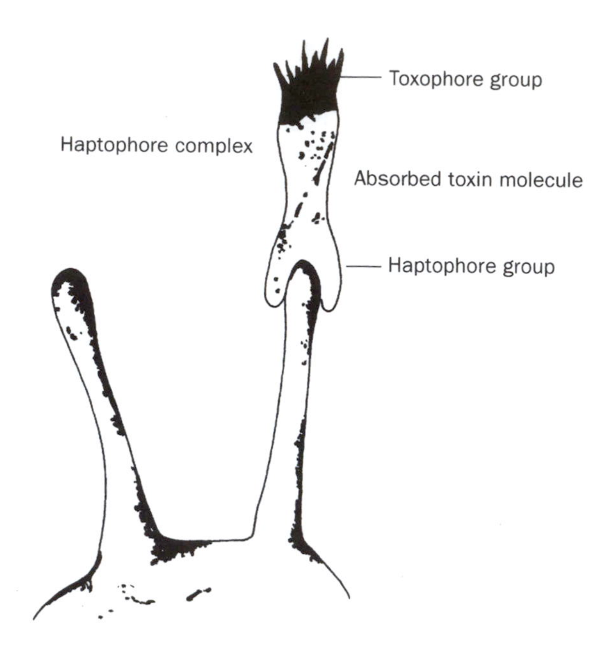

Ehrlich's sketch of a receptor (c. 1898)

and this had been encouraging, in that some amelioration of the symptoms characteristic of the disease had been observed. This positive result seems to have encouraged him to try other dyes on a variety of trypanosomes, the microorganisms responsible for such diseases as African sleeping sickness and Chagas' disease in South America. Laveran and Mesmil, of the Pasteur Institute in Paris, had greatly facilitated such studies when they showed that trypanosomes could be transferred between animals, and these host animals could then be used as a convenient source of the parasites. Ehrlich and his Japanese collaborator Kiyoshi Shiga, showed that Trypan red was effective against trypanosomal infections in mice but had no effect on larger mammals including humans. Their attention was diverted to the use of arsenic compounds when Laveran and Mesnil demonstrated that Trypan red was much more effective if administered in conjunction with arsenious acid. In addition, they became aware of the work of H. W. Thomas of the Liverpool Institute of Hygiene and Tropical Medicine, who had demonstrated the efficacy of the arsenic-containing drug Atoxyl in the treatment of animals infected with trypanosomes. This drug (sodium arsanilate) had been synthesised by Bechamp in 1863 and was about 20 times less toxic than Fowler's solution (mentioned below). Ehrlich and Shiga had earlier tried Atoxyl on cultures of trypanosomes but without success.

That they should have tried arsenic derivatives was not at all unreasonable, since these had been used not only by Paracelsus but also by many physicians in the form of Fowler's solution. The inventor, the Staffordshire physician Thomas Fowler, had extolled the virtues of his solution (mainly potassium arsenite) in 1786 in a report entitled 'Medical Report of the Effects of Arsenic in Cases of Agues, Remittent Fevers and Residual Headaches.' The Scottish missionary and explorer David Livingstone was amongst the first to extoll the virtues of Fowler's solution for alleviation of the symptoms of sleeping sickness. Ehrlich was further influenced in his choice of drug by the studies by, amongst others, his mentor Robert Koch, on the efficacy of the Atoxyl in trypanosomal infections in animals. This particular arsenic derivative had been prepared, as a possible replacement for Fowler's solution, but in 1906, its chemical structure was still not known with certainty, and Ehrlich's first experiments concerned an attempt to confirm its structure. He had received no formal chemical training but seems have to have spent much of his spare time performing test-tube-scale experiments in a laboratory cluttered with unlabelled bottles and jars, and he soon concluded from his observations that the suggested chemical structure was wrong. In particular, it reacted with nitrous acid to form a diazonium salt; hence, he proposed that Atoxyl contained a free amino group. This would make it chemically reactive and would thus be a good starting material for the construction of numerous structural analogues. His chemical collaborators disagreed with him and a furious row ensued . His secretary Martha Marquardt reported the following exchange:

"Atoxyl is *not* an anilide of arsenic acid. On the contrary, it contains a free amino-group... (and in consequence)... . I have deliberately asked for the most simple derivatives to be worked out first. I must ask you to follow my orders strictly. You cannot judge whether this is right or wrong." To which one of the chemists Dr. von Braun replied:

"We cannot accept your directions, and must work according to the classic formula of Bechamp (the discoverer of Atoxyl)." Ehrlich then stormed out of the laboratory with the words: "I adhere to my orders and leave it to you to take the consequences." Two of the three chemists, von Braun and Schmitz, resigned on the spot, but the third, Alfred Bertheim, stayed and worked on the preparation of the compounds decreed by Ehrlich. This research was funded by Casella Dyworks, which was subsequently acquired by the Hoechst Dyeworks. Their synthetic work quickly revealed the accuracy of Ehrlich's proposed structure, and several hundred analogues were prepared and evaluated for biological activity. Compound number 418, later called arsenophenylglycine, for which they proposed a structure containing an arsenic–arsenic double bond was particularly effective against trypanosomes both in laboratory tests and in animals, and was administered to humans after 1907. But the major triumph came when the group moved on to test their compounds against the spirochaete responsible for syphilis, *Treponema pallidum*.

Paul Ehrlich in his laboratory (From the Wellcome Trust Medical Photographic Library)

The origins of syphilis are still a subject of debate, but it seems clear that spirochaetes of the *Treponema* type have always been endemic in Europe and were responsible for a relatively mild form of syphilis. However, the arrival of Columbus and his men in the Caribbean provided the opportunity for exposure of Europeans to the more potent organism *Treponema pallidum*. Their sexual exploits with the local Indians ensured that these adventurers imported the organism into Europe. Here, it almost certainly underwent genetic change to produce a highly contagious and ravaging form of syphilis. This swept Europe at the end of the 15th century in the wake of the marauding French army as it waged war with the Italians. The fall of Naples in 1495 heralded a period of unbridled licentiousness and the inevitable spread of the disease, which is believed to have claimed 10 million lives between 1495 and 1510. The first manifestations of infection were genital sores and the disease usually progressed inexorably to involve the bones and cartilage. The condition was known variously as the 'French disease,' the 'Spanish disease,' the 'English disease' or the 'Great Pox,' and has remained a serious problem, in certain parts of the world, to this day. The original remedy introduced by Paracelsus involved the use of mercury salts and this was still the main treatment well into the 20th century, although the relatively water-insoluble salts like the very toxic mercurous chloride (Calomel) had been largely replaced with the more soluble salts of benzoic acid and salicylic acid, which were less toxic. Mercury compounds can cause all kinds of neurological problems like loss of memory and mental concentration, numbness of the extremities (paraesthesia), trembling and eventually blindness and death. Set against these problems was the desperate need to provide a treatment for a disease that even today has an incidence of around 20 million cases per year, and was then of much greater incidence.

For his new work with *Treponema*, Ehrlich was joined in 1909 by another Japanese collaborator, Sahachiro Hata, an expert on syphilis, who had been the first researcher to succeed in infecting rabbits with the disease. He retested all the arsenic derivatives against *Treponema pallidum* and identified compound 606 as a particularly potent analogue, which possessed curative activity for infected rabbits. Compound 606 had been prepared in 1907, but it had not exhibited any activity against trypanosomes, and Ehrlich subsequently claimed that this had been due to the incompetence of a former collaborator. This lapse and the problems with hypersensitivity reactions seen in some patients receiving arsenophenylglycine persuaded Ehrlich to proceed with great caution. As a result, very extensive animal tests were carried out before 606 was released to selected hospitals for clinical trials.

Ehrlich was well-served by his chosen clinicians, many of whom carried out further animal tests as well as treatment of patients. Excellent results were obtained from hospitals in Pavia, St. Petersburg, Zurich, Altmark and Magdeburg (Germany) and Sarajevo. Samples were also used in England, and the studies at St. Mary's Hospital, Paddington were carried out by a young bacteriologist named Alexander Fleming (soon to be famous for his work on penicillin) and his assistant Leonard Colebrook (later famous for his work on the sulfonamides). In the edition of the *Lancet* of June 17, 1911, they described the use of a new apparatus for the intravenous injection of 606 and gave the following conclusions about its efficacy: "Whether the new drug will displace mercury in the treatment of syphilis remains to be seen… . It, however, has a remarkable effect in causing the lesions to disappear, and especially is this seen in some cases which have resisted mercurial treatment. Much has been made of the dangers inherent in the administration of the drug, but so far as we have gone we have not seen the slightest trace of the evil effects which have been written about in any case which has been injected intravenously."

Encouraged by these successful clinical trials, Ehrlich announced the results at the International Congress for Internal Medicine at Weisbaden on April 19, 1910. The world's press responded to this new treatment with predictable excitement. Ehrlich was inundated with requests for samples of the drug and during the period June to December 1910, an incredible 65,000 doses of the drug were supplied free of charge to any physician who requested it. Not surprisingly, there was some careless administration.

Solutions of the sodium salt of the drug were made with tap water rather than with distilled water, and because the instability of the compound had not been appreciated, many of the samples had decomposed prior to administration. Ehrlich took a great personal interest in these further clinical trials and was much distressed whenever patients suffered as a result of these problems. He subsequently issued very precise instructions for the care and preparation of the drugs.

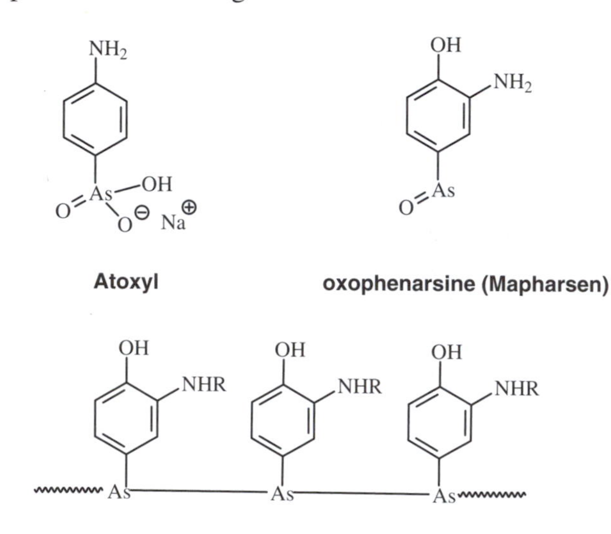

arsphenamine (Salvarsan) R = H
neoarsphenamine (Neosalvarsan) R = $CH_2SO_2^{\ominus}$ $Na^{\oplus}$ or H

(structures probably polymeric rather than with discrete As = As bonds)

The demand for 606 soon outstripped the supply capabilities of Ehrlich's laboratory, and he elicited the support of the Hoechst Chemical Works for the manufacture of the drug, now patented under the name Salvarsan. In contrast to the enthusiastic reception accorded to Salvarsan in Europe, in the USA, the reception was mixed. In America, venereal disease carried a much greater stigma than in Europe. Patients with syphilis were not allowed to be treated in hospitals or dispensaries – the one exception being members of the armed forces. So it was the US Naval Hospital on Mare Island, California, that received the first doses of Salvarsan in October 1910, and 10 patients were treated with at least some dramatic results. As word spread of the efficacy of the drug, some American physicians warned that this remedy might encourage greater sexual promiscuity. One suspects that at least some feared this new quick-acting treatment would reduce the lucrative fees that they received for the long-term treatment of their syphilitic private patients.

The crudity and barbarity of some of the early intravenous injection procedures were often extreme. Instructions for a "one inch incision through the sheath of the vein prior to introduction of the cannula" leave one in little doubt that the procedure was very painful. John Stokes of the Mayo Clinic is said to have warned his physicians to: "Make no comment audible to the patients regarding your needle...and never inject and ask if it hurts, if you have the slightest reason to suspect that it will."

The start of World War I left most European countries without access to a source of Salvarsan. Most countries eventually overturned the German patents and proceeded to manufacture their own Salvarsan and later a more water-soluble version known as Neo-salvarsan (neoarsphenamine). Until the Americans entered the war in 1917, they continued to purchase the drug from the German pharmaceutical industry, although the price rose from $3.50 to around $35 for a single dose. Finally, in October 1917, the US Congress overturned its patent agreements with Germany and the industry began to manufacture 606 under the tradename Arsphenamine. Other arsenic-based drugs were developed following the discovery that Salvarsan was in fact metabolised (that is, chemically altered in the body), and the simpler compound 3-amino-4-hydroxyphenylarsenoxide (oxophenarsine, tradename Mapharsen) entered clinical use in the early 1930's. Mapharsen was more effective and safer than Salvarsan and was still in widespread use at the time of introduction of the penicillins.

None of these drugs fulfilled Ehrlich's vision of a 'magic bullet.' While they all possessed good activity against the spirochaete, that is, they had good parasitotropic activity, they all caused adverse effects in the patient and had some organotropic effects. So how would one judge Ehrlich's overall contributions to chemotherapy? There is no doubt that he demonstrated that drug discovery could proceed through systematic changes in the chemical structure of a compound until an analogue with optimum drug action was obtained: that is, a drug with the best *therapeutic index*. But he also revolutionized the way pharmacologists thought about the way in which a drug interacted with the invading organisms and with the cells of the host (ie, the patient). His early 'side-chain' theory evolved into a much more sophisticated theory concerning *chemoreceptors* for drugs. These were an extension of his 'haptophores' but unlike them they were not released into the bloodstream to act as anti-toxins.

His studies also revealed that after a period of treatment, some trypanosomes or spirochaetes became resistant to the administered drugs, presumably because their chemoreceptors became less specific for the drug. The adverse effects elicited in the patient thus increased as the antimicrobial effect decreased. One solution to this problem was to administer mixtures of drugs with affinities for different classes of chemoreceptors. In this way, smaller quantities of the drugs could be used, and there was an increased chance that the microorganism would be completely eliminated before it had time to become resistant. *Combination chemotherapy*, the mainstay of modern cancer chemotherapy, arose from experiments of this kind.

It would be wrong to ascribe all of these discoveries to Ehrlich. Other scientists were also beginning to talk in terms of chemoreceptors and to design drugs through a process of rational design. For example, the English physiologist John Langley studied the interaction of *inter alia* nicotine, atropine (from deadly nightshade) and curare (the South American arrow poison) on various cells, and concluded in 1905 that: "...there is a chemical combination between the drug and a constituent of the cell – the receptive substance." But Ehrlich was so consistently in the forefront of these developments that he has deservedly come to be known as 'the father of chemotherapy.' His hopes for the future and his realism are perhaps best summarised in his address to the International Congress on Medicine held in London in 1913 (and reported in the *Lancet* of August 16th):

"I believe that now, when definite and secure foundations have been laid for the scientific principles ...of chemotherapeutics, the way is visible before us. In the diseases involving protozoa and spirilla (ie trypanosomal diseases and syphilis) extraordinarily favourable results...have already been gained... . But in contradistinction to these diseases.... the common bacterial diseases (diseases due to streptococcus, staphylococcus, coli, typhoid, dysentery but above all tuberculosis) will require a hard struggle." These words were to prove highly prophetic.

He received numerous prizes and honours, including the Nobel prize for Physiology or Medicine with Metchnikoff (of the Pasteur Institute) in 1908, for his earlier work on immunology. But he always claimed that one of his greatest joys was the election to honorary membership of the German Chemical Society, since this august body only honoured those who had made major discoveries in chemistry. His outstanding contributions to chemotherapy were only just becoming apparent when he died in August 1915. Since then, there have been numerous reviews of his life and work, and even Hollywood was inspired to make a film entitled 'Dr. Ehrlich's Magic Bullet' (in 1940) about the discovery of Salvarsan starring Edward G. Robinson. At the meeting in 1954 to celebrate the centenary of his birth, Sir Henry Dale, himself famous for his work on nerve conduction, said of Ehrlich: "We should remember how large a part of all the great and still expanding advance in the direct and specific treatment of infectious diseases can be traced at its beginning to the inspired vision, the brilliant initiative, and the invincible scientific enthusiasm of Paul Ehrlich" [*Lancet*, March 20th, 1954].

But the final thoughts about Ehrlich should be those of his secretary of 13 years, Martha Marquardt, who saw in him not only a great scientist but a man with enormous humanity: "Nothing on earth mattered to him except scientific research aimed to overcome suffering and disease, and to increase the happiness of mankind. His unshaking and unwavering faith in the progress of his work was like a warming flame which filled his whole nature and shone forth in all his actions" [*Lancet*, March 20th, 1954].

Chapter 2

Fighting Bacteria

THE BLACK DEATH

There can be no better example of the ravages caused by bacteria in a pre-antibiotic world than the loss of life caused by the Great Pestilence of 1347–1352 – usually called the Black Death. This outbreak of plague originated with the black rat (*Rattus rattus*) population of the Gobi Desert in Mongolia in the late 1330s, and the bacillus responsible, *Yersinia pestis*, was passed to the nomadic Mongolian horsemen via the rat flea, most importantly, *Xenopsylla cheopis*. The nomads travelled the caravan routes of Asia and the Middle East, and by 1347 the infection had arrived in the outer reaches of Europe at Astrakhan on the Volga River and at Kaffa on the Black Sea. From there, it was a short trip for the rats and their fleas, courtesy of the Italian merchant fleet, to Constantinople, Alexandria, Messina, Cagliari, Genoa and Marseilles, all of which reported plague deaths in 1347. The arrival in Messina was reported by the friar Michele di Piazze, who wrote:

> In the first days of October 1347, twelve Genoese galleys fleeing before the wrath of our Lord...brought in their bones a disease so violent that whoever spoke a word to them was infected and could in no way save himself from death.

In 1348, the disease raged throughout southern Europe and it also first appeared in England during August of the same year. The probable port of entry for the bacillus was Melcombe Regis, the modern Weymouth, most likely accompanying casks of wine from Bordeaux. The disease spread inexorably northwards and eastwards to Bristol and then to Oxford and reached London in November of that year. Between 1349 and 1352, virtually the whole of northern Europe, including parts of Scandinavia and Russia, was in the grip of the plague, and it has been estimated that around a quarter (perhaps as much as a third) of the population died as a result of infection.

Three forms of the disease were recognised: pneumonic plague, septicaemic plague and, most common of all, bubonic plague, spread by the bites of rat fleas. Bubonic plague was characterised by grossly swollen lymph nodes, usually in the groin or neck. These were called buboes (from the Greek *boubon* meaning groin), and they produced excruciating pain before rupturing. In septicaemic plague, there was an extreme form of infection and the sufferer usually died before the buboes appeared. If the bacillus invaded the lungs, these patients could transmit their infection by coughing, and this led to pneumonic plague in those who inhaled the droplets of bacillus. Bubonic plague resulted in a death rate of around 30–75%, whilst the other two forms proved to be rapidly fatal in almost 100% of those affected.

Boccaccio in the *Decameron* presents the stories of seven persons who fled the city of Florence to avoid contracting the plague, and provided a graphic account of the symptoms and the response of the populace to this manifestation of God's displeasure:

> the appearance of swellings in the groin and armpit, some of which were like a common apple and others were egg-shaped [*come una communal mela ed altre come una uova*].

He wrote that the people called these swellings *gavoccioli* and that they usually spread to produce dark bruises on the arms, thighs and other parts of the body, and that this was a sure sign of their imminent demise: "*era certissimo indizio di futura morte*."

The response of the populace to the arrival of the plague varied from the adoption of a sober and abstemious lifestyle to ensure God's forgiveness, to the opposite extreme where they believed that they might as well eat, drink and be merry: "to satisfy all of their appetites for wild living" [*il sodisfare d'ogni cosa all'appetito che si potese*].

Many cities initiated a rather belated attempt to clear the rotting rubbish and sewage from the streets, or to enforce strict quarantine rules (from the Italian *quarantena*, meaning to hold for 40 days). In Milan, this included boarding up the houses where plague had been reported, with those infected still inside, leaving them to die. Inevitably, Jews, Muslims and lepers were held to blame in some centres, with a wholesale slaughter of Jews in several European cities. However, most people believed that the disease was caused by *miasmata* or evil vapours, and they moved around with posies of fragrant herbs and flowers clutched to their noses.

The population of England was probably around four million at the start of the plague and it is likely that at least one million perished during the four-year period. The loss of life in the cities was proportionately much higher than in the relatively healthy environment of the countryside, and since England's economy was still largely dependent upon its peasant farmers, this prevented the total collapse of the economy. However, the long-term social and demographic trends were quite marked. In Florence, for example, the city from which Boccaccio fled, the population of more than 100,000 declined by at least two-thirds during the 90-year period that began in 1348. It is worth noting, however, that while the plague was responsible for sudden increases in the annual death rates during plague years, other diseases caused a smaller but more regular loss of life over many centuries. For example, each year during the winter months, there was a peak in the number of deaths and this was due to typhus, smallpox and respiratory infections. During the early autumn of each year, a similar increase in the death rate was undoubtedly caused by diphtheria, measles and dysentry. There was one other significant effect of this first plague, and that was the enforced cessation of hostilities between the English and French following the battle at Crécy in 1346. There were simply not enough able-bodied troops available during the plague years, although hostilities resumed in 1355 once the plague began to abate.

From 1350 to the late 17th century, the plague was endemic in Europe. A further catastrophe occurred in London between the years 1504 and 1505, when perhaps one-third of the population (of 60,000) of London died as a result of the Black Death. Other major epidemics occurred between 1558 and 1564, in 1603, 1625 and, most importantly, and for the last time, in 1665 – 1666. This final outbreak of the plague may have caused 100,000 deaths (out of a population of 500,000) in London, and is described by Samuel Pepys in his diary. The disease had been essentially dormant (like the rats) during the long, cold winter of 1664–1665, but re-emerged with a vengeance during the heat wave of summer 1666. On June 7, Pepys wrote in his diary: "It being the hottest day that ever I felt in my life, and it is confessed so by all other people."

On June 20 he commented that "there died four or five at Westminster of the plague" and by August 12 he showed considerably more concern:

> The people die so, that now seems they are fain to carry the dead to be buried by daylight, the nights not sufficing to do it in. And my Lord Mayor commands people to be within at 9 at night...that the sick may have liberty to go abroad for ayre.

Pepys and his wife then left London for the relative safety of the countryside, but he returned for a daytrip on October 16 and wrote:

> But Lord, how empty the [city] streets are, and melancholy, so many sick people in the streets, full of sores... . And they tell me that in Westminster there is never a physitian, and but one apothecary left, all being dead.

The plague abated during the winter and the Great Fire, which started in the early hours of September 2, 1666, drove out the rat population and allowed the city to be rebuilt with more attention given to a clean water supply and the disposal of refuse and sewage. Similar efforts elsewhere in Europe led to the complete eradication of the bacillus from the human population, although it continued to cause sporadic outbreaks of plague in other parts of the world right up to the present time, with the most recent one being in India in 1994. The main reservoir for the bacillus is the wild rodent population, and it is found in more than 200 different species, with transmission by at least 30 species of fleas. Most instances of the disease now occur if people disturb the natural habitat of these rodent species.

Bubonic plague no longer causes the great dread that it once did, because there are now several very effective antibacterial drugs that can destroy the bacillus. These include streptomycin, chloramphenicol and the tetracyclines, but not the penicillins or related drugs. Nonetheless, the sheer scale of the loss of life in the pre-antibiotic era provides a timely reminder of how much we have grown to depend upon these 'magic bullets'.

Miasmata, animalcules and *mal aria*

The agents responsible for these great losses of life were a complete mystery. From ancient times, people had speculated about emanations (*miasmata*) of disease arising from swampy places (like mist) and from corpses. This bad air or *mal aria* was associated with the fevers and worse conditions that were prevalent in marshy places. In 1546, the Italian physician and poet Girolamo Fracastoro proposed what we should now term the 'germ theory of disease' when he wrote in his treatise entitled *De Contagione*:

> ...all infections may be reduced ultimately to putrefaction...and, if all infection is putrefaction, infection in the ordinary sense...is nothing else than the passage of a putrefaction from one body to another.

His theory was, however, ridiculed. Others, like the English physicist Robert Boyle, anticipated the importance of fermentation that would be finally established by Pasteur in the late 19th century. Writing in 1663, Boyle stated:

> He that thoroughly understands the nature of ferments and fermentation, shall probably be better able...to give a fair account of divers phenomena of several diseases... .

Ten years later, Antoni van Leeuwenhoek examined many natural substances, including scrapings from his teeth and his faeces under the lens of his new microscope and described the myriad *animalcules* that we should now call protozoa, fungi and bacteria. There remained, however, a firm belief that such creatures were the product of spontaneous generation and this was only disproved by Pasteur's experiments on fermentation.

Louis Pasteur was born in Dole in the Jura in 1822. After receiving his doctoral degree from the Ecole Normale Superieur in Paris in 1847, he was appointed, in turn, Professor of Physics in Dijon (1848), of Chemistry in Strasbourg (1849) and Lille (1854), and then returned to the Ecole Normale Superieur as Director of Scientific Studies in 1857, before his appointment as Professor of Chemistry at the Sorbonne in 1867. His long and successful research career began in 1848 with his discovery of optical isomers. Using a microscope, Pasteur was able to identify two mirror image forms of sodium

Louis Pasteur dictating a research paper to his wife

ammonium tartrate crystals. He quickly realised that many other naturally occurring substances existed in two discrete forms, and much more critically, only one of these forms could be used by living systems. This probably prepared his mind for subsequent discoveries with ferments (where the specificity is due to enzyme activity).

Pasteur's studies with microorganisms commenced around 1855 and were initially confined to the fermentation associated with beer and wine production, and the formation of lactic acid by the souring of milk. According to the great French chemist Lavoisier, whose brilliant career was abruptly terminated by the guillotine, the fermentation of glucose to produce alcohol involved the splitting of the molecule into two parts: an oxidised product – carbon dioxide – and another product – alcohol – that had lost oxygen (been reduced). This theory, however, did not ascribe any function to the yeast.

Pasteur's study on the conversion of sugar to lactic acid was reported in his *Memoire sur la fermentation appelée lactique* and appeared to show that a yeast-like material was involved. His most significant demonstration was that the 'ferment' could be grown in a nutrient solution where it multiplied to produce a population of microscopic organisms. In his later studies on alcohol production, summarised in his *Memoire sur la fermentation alcoholique*, published in 1857 (and in more detail in 1860), he showed that living organisms were involved. In the book, he noted:

> Alcoholic fermentation is an act correlated with the life and with the organisation of these globules (yeast cells), and not with their death or their putrefaction.

He went on to show that certain fermentations would only proceed in the absence of oxygen and coined the terms *anaerobic* (for such processes) and *aerobic* for those that required oxygen. Indeed, much of his success can be ascribed to the fact that he recognised that efficient sterilization by heat treatment and subsequent exclusion of oxygen (in the air) could lead to the preservation of beer and milk because fermentation failed to occur. These discoveries have been immortalised by use of the term *pasteurisation* to describe this method of preservation. All these studies allowed him to propose the central theory that dominated the rest of his studies, namely that transformation of organic matter is effected by specific microorganisms with particular requirements for nutrients and temperature if they are to have optimum activity.

Unknown to Pasteur, a young German doctor by the name of Robert Koch, working in the country town of Wollstein (now part of Poland), was also studying the actions of microorganisms. Koch was born in Clausthal near Hannover in December 1843, and after obtaining his MD from the University of Göttingen in 1866, he worked as an army surgeon during the

Franco-Prussian War of 1870–1871. After the war, he became particularly interested in understanding how animals and humans became infected with anthrax. This was, at that time, a relatively common disease of domestic animals, and since the bacilli of *Bacillus anthracis* were relatively large, they could be easily observed in the blood of infected animals even with the primitive microscopes of the time. In the early part of the 19th century, various investigators had observed what they believed to be anthrax microorganisms, but had failed to appreciate or demonstrate their infectious nature. It was Robert Koch who showed, in 1876, that it was possible to grow the bacilli in a culture containing the aqueous humour Ox eye, and that after several rounds of sub-culturing, the organisms were still just as infectious as the original sample. Sceptics, however, raised the objection that the aqueous humour could be carrying the infection from culture to culture.

Pasteur was by now aware of these studies and was able to overcome this scepticism by some clever experimentation. He added one drop of infected serum to sterilised urine, allowed the bacilli to multiply and then added one drop of this urine to another sample of sterilised urine. After 100 transfers, the anthrax bacilli retained the entire potency present in the initial drop of blood serum, and the germ theory of disease was thus confirmed.

In the meantime, Robert Koch had identified the microorganisms responsible for tuberculosis (*Mycobacterium tuberculosis*) in 1882 and cholera (*Vibrio cholerae*) in 1883, and went on to propose several rules for determining if a disease is caused by a microorganism. These became known as *Koch's postulates*, and although they now appear rather obvious, they were revolutionary at the time of their proposal. In essence, they state that the organism must be present whenever the disease is diagnosed, and after incubation in a culture or in another animal, the same organism can be isolated from the newly infected culture or animal. Koch's work received international acclaim and he became Professor and Director of the Institute of Hygiene at the University of Berlin in 1885. Somewhat belatedly, he received the Nobel Prize for Physiology or Medicine in 1905, having spent the later years of his research career studying tropical diseases. He died in May 1910.

Koch had also experimented with dyestuffs as a means of helping with the identification of microorganisms and this eventually led to the division of bacteria into two main classes. These are the Gram-positive organisms like streptococci (responsible for pneumonia, ear and throat infections, *etc*.), staphylococci (causing boils, endocarditis, meningitis, *etc*.), pneumococci (causing pneumonia), and Gram-negative organisms like *Salmonella typhimurium* (which causes typhoid fever) and *Escherichia coli* (causing food poisoning, urinary tract infections, *etc*.). The Gram-positive organisms will react irreversibly with the stain crystal violet and this colouration resists washing and reaction with other dyestuffs, and they thus appear violet. The

Gram-negative organisms also interact with methyl violet, but the colouration can be removed by washing and the bacteria can then take up a different stain, usually safranine, which stains them pink. This classification is named after Hans Christian Joachim Gram, a Danish bacteriologist, who first described it in 1884. It is, of course, an extension of the ideas of Ehrlich who first used a dyestuff to identify the tubercle bacillus.

All bacteria possess a relatively rigid cell wall outside of an inner cell membrane. The light microscope easily reveals that they exist in several main forms: those that are essentially spherical in shape (*cocci*); those that appear as short, straight cylinders (*bacilli*) or slightly curved cylinders (*vibrios*); and those that occur as spiral rods (*spirilla*). The different interactions with the Gram stain largely reflect the chemical constitution of the cell wall. Gram-positive organisms have a relatively thick wall (between 15 and 80 nm), which comprises a mixture of glycopeptides (peptidoglycan) and teichoic acids, which are polymers based on ribitol and glycerol linked by phosphate units. The glycopeptides typically comprise a polysaccharide composed of repeating units of *N*-acetlyglucosamine linked to *N*-acetylmuramic acid, with a polypeptide chain attached to the lactic acid moiety of *N*-acetylmuramic acid. In contrast, Gram-negative organisms have a much thinner wall, which comprises a peptidoglycan of 3 nm thickness surrounded by an outer membrane (8 nm) composed of a complex mixture of protein, phospholipid and lipopolysaccharide. These complex polysaccharides have antigenic properties and their large variety means that Gram-negative bacteria exist in numerous serotypic forms – salmonella alone has more than 1000 serotypes.

Within the bacterial cell is the cytoplasm where most of the biochemistry takes place. The cell's DNA is not confined to a nucleus as in cells of higher organisms, but its replication to produce various types of RNA and thence all the bacterial enzymes and proteins, proceeds in much the same way as in higher organisms. The nutrients and other chemical species required for these processes pass into the cell via the cell membrane. They usually pass unimpeded through the cell wall material, but the membrane does offer a barrier with selective permeability for those chemicals that are required for the biochemistry. Most antibacterial drugs interfere in some way with either the biochemical processes occurring in the cytoplasm or with those processes that produce the cell wall material. The best drugs have a highly specific effect on these biochemical reactions, but the older antibacterial agents, like Lister's carbolic acid, were generally toxic to bacterial cells.

ANTISEPTICS

The ancient Egyptians knew a thing or two about putrefaction and how to prevent it. Their embalming mixtures comprised aromatic plant extracts, tree resins, pitch and spices, and the abundance of well-preserved mummies in our museums provides cogent proof of their efficacy. The famous Greek physician Hippocrates, in the 5th century BC, used tar and wine on wound dressings. In the 16th century, the noted military surgeon Ambroise Paré used a decoction made from sage, rosemary, thyme, lavender, flowers of camomile and meliot, red roses boiled in white wine, and a dessicant made of oak ashes, a little vinegar and half a handful of salt. This he claimed would 'attenuate, incise, resolve, wither and dry up the thick, viscous humour and prevent the transfer of vapours to the heart and brain'. This 'viscous humour' is probably a reference to what was generally known as 'laudable pus', which was widely believed to be an essential part of the healing process rather than, as we now know, a product of infection. Many of the chemical compounds present in these various preparations were of two types: phenols and terpenes, both of which have antimicrobial activity. There was a practice, in parts of Yorkshire, of saving part of the Easter cake to allow it to grow mould, which was then scraped off and used as an antiseptic ointment. Since the mould may well have been of the *Penicillium* family, these country folk could well have unwittingly been using a crude mixture of penicillins.

No systematic search for antimicrobials was carried out until the 18th century when Sir John Pringle, a military surgeon, carried out a series of experiments between 1750 and 1752 to determine the most efficacious way of retarding the putrefaction of beef. He described most of his results in a series of papers published by the Royal Society in 1750. Amongst the common chemicals he studied were salt, alum, borax, nitre (potassium nitrate), salt of hartshorn (ammonium carbonate), sal ammoniac (ammonium chloride) and the monoterpene camphor. Camphor was the best agent for retarding spoilage, with alum a close second, and he used the term *antiseptic* to describe such chemicals. The first use of this word is, however, usually ascribed to a person by the name of Mr. Place, who used the term in a pamphlet entitled *An hypothetical notion of the plague and some other of the many thoughts about it* (1721). Sir John Pringle was more famous for his work to improve hygiene in the army, and his attempts to eradicate 'jail fever' and 'hospital fever', which were, in reality, typhus. In addition, he championed the idea that army hospitals should be recognised as sanctuaries, especially in times of war. The horrors endured by the wounded at the Battle of Solferino (June 24, 1859) were witnessed by the Swiss banker Jean Henri Dunant, who was inspired to bring about the establishment of the Red Cross in 1864.

The Swedish chemist Carl Scheele first isolated chlorine in 1774 and the bleaching properties of its aqueous solutions were discovered soon afterwards. Various preparations were marketed for this purpose, but its medical use was not explored until 1825 when the Parisian apothecary Labarraque began to sell Eau-de-Labarraque as a wound disinfectant. But the major advance originated from attempts to prevent the large number of deaths from puerperal fever. This was a common sequel to childbirth, especially when women gave birth in hospitals rather than in their own homes. In this environment, they were likely to encounter doctors (and perhaps midwives) whose attention to cleanliness was non-existent, and as a result uterine infection by the bacterium *Streptococcus pyogenes* was very prevalent. In some hospitals, the death rate was as high as one in five of newly delivered women. As early as 1795, the Scottish surgeon Alexander Gordon had drawn attention to the possibility that poor hygiene was the cause of puerperal fever. His views were substantiated by the studies of Oliver Wendell-Holmes in Boston and Ignaz Semmelweis in Vienna in the mid-1800s. Semmelweiss observed that the death rate in the wards where medical students were trained (on cadavers) were far higher than in wards exclusively for the use of pre-natal patients. Both men insisted that their respective hospitals were disinfected with calcium hypochlorite (obtained by the passage of chlorine gas into calcium hydroxide solution), and succeeded in reducing the incidence of puerperal fever deaths from around 18% to almost zero. Both of them received much abuse from colleagues who resented their assertions of uncleanliness; Semmelweiss eventually went insane and ultimately died from septicaemia after pricking his finger.

However, the efficacy of their disinfection programmes encouraged other hospitals to adopt the practices. Other antiseptic preparations rose in popularity towards the end of the 19th century, most notably, various chloramines like chloramine T and halazone (which slowly release hypochlorous acid), tincture of iodine and iodoform; but it was the intro-

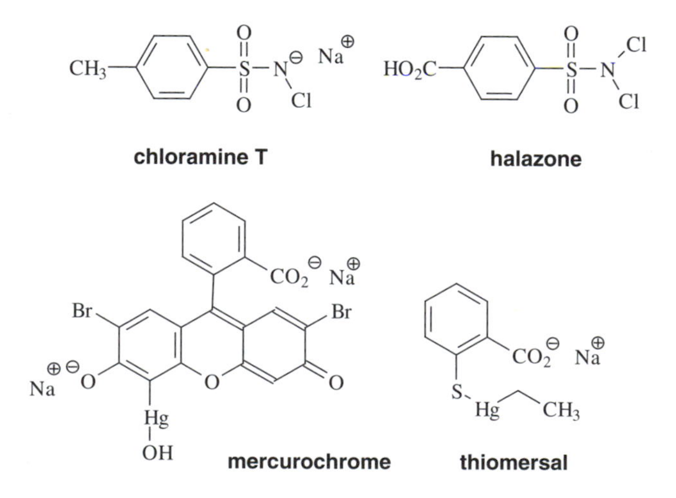

duction of carbolic acid by Joseph Lister that revolutionised the prevention of infection in hospitals.

The modern petrochemical industry had its origins in the Industrial Revolution, although it was the chemistry of coal rather than oil that was explored. During the 19th century, various products from coal tar had been used as disinfectants for water supplies; but it was not until 1833 that Friedlieb Runge isolated pure carbolic acid by distillation of coal tar. He was, incidentally, also the first to isolate, in 1819, quinine from the bark of the cinchona tree, and thus provided the basis for the wide-scale prophylaxis for malaria. Although he reported the preservative properties of carbolic acid, it was not commercially available until 1857, when the Manchester chemist Frederick Calvert devised a means of producing it on a large scale, and persuaded several local authorities to use it for the treatment of sewage.

Josesph Lister was born in April 1827 in East London close to what is now West Ham. His father was a prosperous wine merchant; hence, he benefited from a good education at the Quaker School in Tottenham. He seems to have had a strong interest in natural history and anatomy and had probably decided to be a surgeon before leaving school. His Quaker religion denied him access to Oxford and Cambridge, and he enrolled for his university education at the newly instituted University College known as the 'Godless Institution on Gower Street'. His father advised him to study for an Arts' degree, and he obtained his BA in 1847, but also commenced his training to be a surgeon. While at University College, he was privileged to be present when ether anaesthesia was used for the first time in December 1846 – chloroform was first used at Edinburgh Royal Infirmary in November 1847. After completion of his training in 1854, he was fortunate to obtain a post as resident house surgeon in the surgical team led by James Syme at the Edinburgh Royal Infirmary. At that time, Syme was probably the most famous surgeon in Britain (and probably Europe too), although the facilities available to him in the hospital left much to be desired. The 72 beds in his six wards were under the care of two staff nurses and seven nurses. One of Lister's colleagues, Joseph Bell, described these nurses as "poor useless drudges, half charwoman, half field-worker, rarely keeping their places for any length of time, absolutely ignorant, almost invariably drunken, sometimes deaf and often fatuous."

Lister made good progress and was appointed Professor of Surgery at the University of Glasgow in 1860, and spent the next five years urgently seeking an agent that would prevent the almost inevitable sepsis that followed major (and even minor) surgery. Localised sepsis very often led to generalised septiceamia and death in a high proportion of cases, and there was also a high incidence of tetanus (lockjaw) caused by the organism *Clostridium tetani*, erysipelas caused by group A streptococci, and what was known as hospital gangrene, of uncertain origins. This last, ill-defined condition led to

Lord Lister (Photo from the Royal College of Surgeons, Edinburgh)

the death of as many as 25% of all surgical patients. The high rate of infection is hardly surprising since the surgeons usually operated in street clothes and their instruments received only the minimum of cleaning prior to use.

It is clear that Lister understood the significance of Pasteur's work on microorganisms. In his *British Medical Journal* paper of 1867 (vol. ii, p. 246), he wrote:

> But when it had been shown by the researches of Pasteur that the septic property of the atmosphere depended, not on oxygen or any gaseous constituents, but on minute organisms suspended in it... it occurred to me that decomposition in the injured part might be avoided without exclusion of

> air, by applying as a dressing some material capable of destroying the life of the floating particles.

Lister acknowledged his debt to Pasteur's discoveries in a letter he wrote to him in 1874:

> Allow me to take this opportunity to tender my most cordial thanks for having, by your brilliant researches, demonstrated to me the truth of the germ theory of putrefaction, and thus furnished me with the principle upon which alone the antiseptic system can be carried out.

Lister was also aware that the local authority in Carlisle was using (at Calvert's instigation) carbolic acid for sewage treatment, but was unable to obtain any supplies until 1865. He then treated an 11-year-old boy with a compound fracture of the leg (he had been run over by a cart). Such accidents were common and this type of fracture often proved fatal due to subsequent septicaemia. Lister set the bones and covered the open wound with a dressing soaked in carbolic acid and linseed oil, and the boy survived. Unaware that similar studies had been carried out by the French surgeon Jules Lemaire in 1863, Lister wrote to the *Lancet* describing his success in March 1867. He treated a further ten cases during the next month and there was only one fatality. This was due to a sudden haemorrhage rather than to post-operative infection.

Despite this and other successes, envious colleagues refused to accept his discoveries. However, the general public became very enthusiastic about this new 'wonder chemical', and use of such preparations as the 'carbolic smoke ball' became very popular. The carbolic acid was placed in a heat-proof vessel and this was warmed so that the vapours could be inhaled. Anyone who has encountered carbolic acid (phenol) would doubt that this fairly corrosive chemical could provide the benefits claimed in the advertisement!

In 1867, Lister introduced the use of carbolic acid-treated ligatures and then catgut (sheep's intestine) similarly treated, with great success. At the same time, he earned the wrath of the managers of the Glasgow Royal Infirmary when he described the hospital wards as "the most unhealthy in the kingdom". Despite his entreaties that he had meant to imply that the improved health of his patients had been achieved through antiseptic practices rather than by other factors like improved nutrition, he was not easily forgiven. Nonetheless, the recorded mortality figures for his amputations prior to 1867 and the use of carbolic acid (>45%) fell dramatically (to around 15%) between 1867 and 1870, providing cogent support for his methods. Some idea of the dangers of amputation without the use of antiseptics can be gleaned from the appalling statistics associated with the battles of the Franco-Prussian war in 1870–1871, when more than 13,000 amputations were carried out with greater than 10,000 resultant deaths from gangrene and other bacteria-related causes.

JAN. 23 1892 THE ILLUSTRATED LONDON NEWS

CARBOLIC SMOKE BALL

WILL POSITIVELY CURE

COUGHS Cured in 1 week
COLD IN THE HEAD Cured in 12 hours.
COLD ON THE CHEST Cured in 12 hours.
CATARRH Cured in 1 to 3 months.
ASTHMA Relieved in 10 minutes.
BRONCHITIS Cured in every case.
HOARSENESS Cured in 12 hours.
LOSS OF VOICE Fully restored.
SORE THROAT Cured in 12 hours.
THROAT DEAFNESS Cured in 1 to 3 months.
SNORING Cured in 1 week.
SORE EYES Cured in 2 weeks.
INFLUENZA Cured in 24 hours.
HAY FEVER Cured in every case.
HEADACHE Cured in 10 minutes.
CROUP Relieved in 5 minutes.
WHOOPING COUGH Relieved the first application.
NEURALGIA Cured in 10 minutes.

As all the Diseases mentioned above proceed from one cause, they can be Cured by this Remedy.

£100 REWARD

WILL BE PAID BY THE
CARBOLIC SMOKE BALL CO.
to any Person who contracts the Increasing Epidemic,

INFLUENZA,

Colds, or any Diseases caused by taking Cold, after having used the **CARBOLIC SMOKE BALL** according to the printed directions supplied with each Ball.

£1000 IS DEPOSITED

with the ALLIANCE BANK, Regent Street, showing our sincerity in the matter.

During the last epidemic of **INFLUENZA** many thousand **CARBOLIC SMOKE BALLS** were sold as preventives against this disease, and in no ascertained case was the disease contracted by those using the **CARBOLIC SMOKE BALL.**

Free Trials at our Consulting Rooms.
For Inhalation Only.

Free Trials at our Consulting Rooms.
For Inhalation Only.

THE CARBOLIC SMOKE BALL,

TESTIMONIALS.

The DUKE OF PORTLAND writes: "I am much obliged for the Carbolic Smoke Ball which you have sent me, and which I find most efficacious."

SIR FREDERICK MILNER, Bart., M.P., writes from Nice, March 7, 1890: "Lady Milner and my children have derived much benefit from the Carbolic Smoke Ball."

Lady MOSTYN writes from Carshalton, Cary Crescent, Torquay, Jan. 10, 1890: "Lady Mostyn believes the Carbolic Smoke Ball to be a certain check and a cure for a cold, and will have great pleasure in recommending it to her friends. Lady Mostyn hopes the Carbolic Smoke Ball will have all the success its merits deserve."

Lady ERSKINE writes from Spratton Hall, Northampton, Jan. 1, 1890: "Lady Erskine is pleased to say that the Carbolic Smoke Ball has given every satisfaction; she considers it a very good invention."

Mrs. GLADSTONE writes: "She finds the Carbolic Smoke Ball has done her a great deal of good."

Madame ADELINA PATTI writes: "Madame Patti has found the Carbolic Smoke Ball very beneficial, and the only thing that would enable her to rest well at night when having a severe cold."

AS PRESCRIBED BY
SIR MORELL MACKENZIE, M.D.,
HAS BEEN SUPPLIED TO
H.I.M. THE GERMAN EMPRESS.

H.R.H. The Duke of Edinburgh, K.G.
H.R.H. The Duke of Connaught, K.G.
The Duke of Fife, K.T.
The Marquis of Salisbury, K.G.
The Duke of Argyll, K.T.
The Duke of Westminster, K.G.
The Duke of Richmond and Gordon, K.G.
The Duke of Manchester.
The Duke of Newcastle.
The Duke of Norfolk.
The Duke of Rutland, K.G.
The Duke of Wellington.
The Marquis of Ripon, K.G.
The Earl of Derby, K.G.
Earl Spencer, K.G.
The Lord Chancellor.
The Lord Chief Justice.
Lord Tennyson.

TESTIMONIALS.

The BISHOP OF LONDON writes: "The Carbolic Smoke Ball has benefited me greatly."

The MARCHIONESS DE SAIN writes from Padworth House, Reading, Jan. 13, 1890: "The Marchioness de Sain has daily used the Smoke Ball since the commencement of the epidemic of Influenza, and has not taken the Influenza, although surrounded by those suffering from it."

Dr. J. RUSSELL HARRIS, M.D., writes from 6, Adam Street, Adelphi, Sept. 24, 1891: "Many obstinate cases of post-nasal catarrh, which have resisted other treatment, have yielded to your Carbolic Smoke Ball."

A. GIBBONS, Esq., Editor of the *Lady's Pictorial*, writes from 172, Strand, W.C., Feb. 14, 1890: "During a recent sharp attack of the prevailing epidemic I had none of the unpleasant and dangerous catarrh and bronchial symptoms. I attribute this entirely to the use of the Carbolic Smoke Ball."

The Rev. Dr. CHICHESTER A. W. READE, LL.D., D.C.L., writes from Banstead Downs, Surrey, May 1890: "My duties in a large public institution have brought me daily, during the recent epidemic of influenza, in close contact with the disease. I have been perfectly free from any symptom by having the Smoke Ball always handy. It has also wonderfully improved my voice for speaking and singing."

The Originals of these Testimonials may be seen at our Consulting Rooms, with hundreds of others.

One CARBOLIC SMOKE BALL will last a family several months, making it the cheapest remedy in the world at the price—10s., post free.

The CARBOLIC SMOKE BALL can be refilled, when empty, at a cost of 5s., post free. Address:

CARBOLIC SMOKE BALL CO., 27, PRINCES ST., HANOVER SQ., LONDON, W.

The carbolic smoke ball

In 1870, Lister moved on to the Chair of Clinical Surgery at Edinburgh Royal Infirmary, and it was there that the carbolic acid spray was invented. This contained a mixture of one part of carbolic acid and 100 parts of water, and this mixture was sprayed all around the patient and the surgical staff during the operation. Not surprisingly, the surgical staff suffered from what was termed 'carbolic hands', which involved sores and blisters; but Lister's methods became ever more popular, especially in Europe and the USA. In 1877,

The dangers of amputation (From the Wellcome Trust Medical Photographic Library)

he returned to London to take up the Chair of Surgery at King's College Hospital and here he experimented with other potential antiseptics, including thymol, eucalyptus extract and salicylic acid. The latter was already proving to be successful as an anti-inflammatory agent and was the forerunner of aspirin, but none of these alternative chemicals was a good antiseptic. Around this time, Koch pointed out that although carbolic acid could inhibit the growth of bacteria, it did not kill them. Indeed, this was true of most chemical substances, and of the ones he tested, only mercuric chloride could be guaranteed to kill most types of bacteria. Lister accepted these findings, and eventually discovered that the mixed cyanides of mercury and zinc in combination with the dyestuff mauveine were extremely effective, and this preparation gradually replaced carbolic acid in hospitals. This was not the only preparation based on mercury that became widely used, for some years earlier Robert Koch had pioneered the use of mercuric chloride (first popularised by Paracelsus). This led on to the discovery of mercury benzoate and salicylate for the treatment of syphilis, and thence to mercurochrome, thiomersal, *etc.*, as topical antiseptic agents. (The latter is used as a preservative for certain polio vaccines, and was at the centre of a recent frenzy of media disinformation when it was reported that toxic 'mercury' was present in these vaccines. At the time, I was asked by a local television station if I could

The carbolic acid spray generator

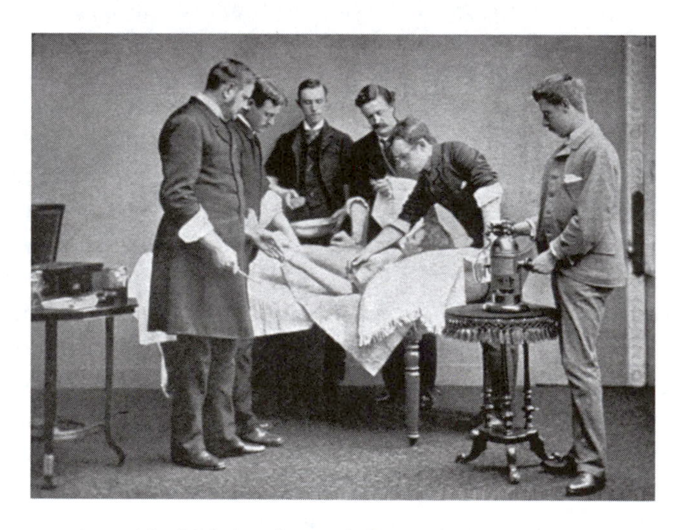

Lord Lister operating with the help of a carbolic spray generator (Photo from the Royal College of Surgeons, Edinburgh)

arrange a laboratory demonstration of mercury being poured into a beaker for one of their news items. I declined, although I did try to explain the difference between a mercury compound and elemental mercury, but to no avail since they ran the piece about 'toxic mercury' anyway!)

As his fame spread, numerous honours were bestowed on Lister. He became surgeon to Queen Victoria in 1878, although he had already operated on her (for an abscess) in 1870 whilst she was resident at Balmoral. On this occasion, she had complained about the pungency of the carbolic acid, but eventually concluded that the operation had been "a most disagreeable duty most pleasantly performed." He became a baronet in 1893 and a peer in 1897. But one of the most memorable and historically significant events in which he participated was the celebration to mark the 70th birthday of Louis Pasteur. This was held at the Sorbonne in Paris, and Lister was invited to attend as the representative of the Royal Societies of London and Edinburgh. In his speech, Lister paid tribute to Pasteur's research that had "raised the veil which had covered infectious diseases for centuries, and had discovered and demonstrated their microbial nature". In his reply, Pasteur spoke (his son actually read the speech) of his belief that:

> Science and peace will triumph over ignorance and war, nations will unite, not to destroy, but to build, and the future will belong to those who will have done most for suffering humanity.

These words provided a fitting tribute to all the pioneers of bacterial research. The fundamental knowledge was now in place so that the search for effective antibacterial agents could begin in earnest.

FROM DYES TO SULFONAMIDES: THE FIRST ANTIBACTERIALS

Lister's carbolic acid may have revolutionised surgical practice, but it could not be administered to patients in order to cure bacterial infection because of its corrosive nature. During the first years of the 20th century, a few derivatives of the well-established antimalarial drug, quinine, were prepared by the German chemist Julius Morgenroth in Frankfurt, and evaluated for antibacterial activity. Contemporaneously, the Glaswegian chemist Carl Browning showed that certain acridine dyes had broad-spectrum antibacterial activity, and one of these, proflavine, eventually entered clinical use during the Second World War. Interestingly, both these pioneers had been students with Paul Ehrlich. Hundreds of acridine dyes were prepared by the Hoechst Dyeworks in Elberfield during the 1920s, and although none of these had useful antibacterial properties, the antimalarial drug mepacrine eventually evolved from this programme of research.

One of the problems with these early endeavours was the lack of reproducible and reliable antibacterial tests. With a view to extending the search for broad-spectrum antibacterial agents, I.G. Farbenindustrie in Elberfield (also owners of the Hoechst Dyeworks) appointed Gerhard Domagk to screen for such activity. His decision to employ mice that were infected with *Streptococcus pyogenes* for antibacterial screening was a brilliant move. This virulent organism was relatively common and had been responsible for many of the deaths in the great influenza pandemic of 1918–1919 (discussed in detail in the next chapter). It was also frequently involved in the causation of bacterial meningitis, rheumatic fever and puerperal fever. The strain he cultured was, in fact, taken from a patient who had died from septicaemia.

Initially his research group studied acridine dyes and gold compounds, but the major breakthrough came when they studied the antibacterial effects of azo dyes. Their first success came with a relatively simple derivative of chrysoidine, which had a side-chain structure similar to that of mepacrine, and this had good activity against streptococci in culture but not in living mice. Introduction of a sulfonamido group into the molecules produced several dyes that had antibacterial activity in mice but not against bacteria grown in culture. One of these dyes, Prontosil Rubrum or Streptozon, proved to be particularly potent. When he administered the dye via a tube directly into the stomachs of 12 mice infected with streptococci, all of them survived. Fourteen control animals (infected but not treated) died.

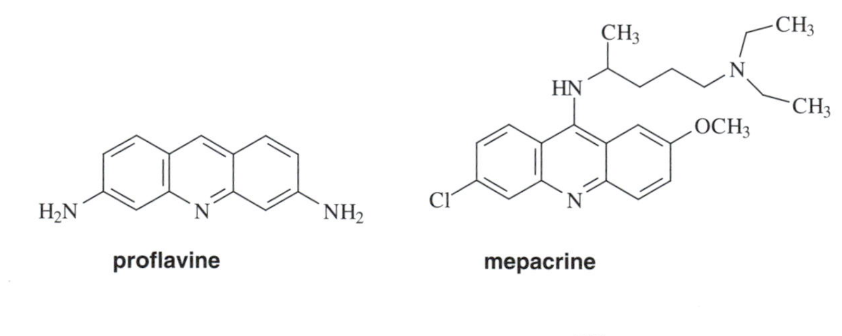

proflavine **mepacrine**

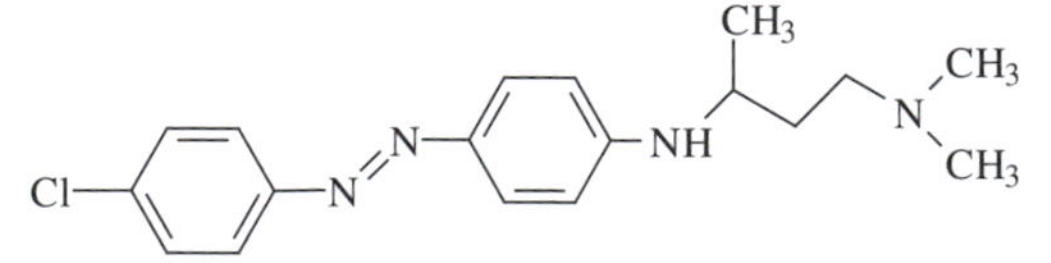

chrysoidine analogue

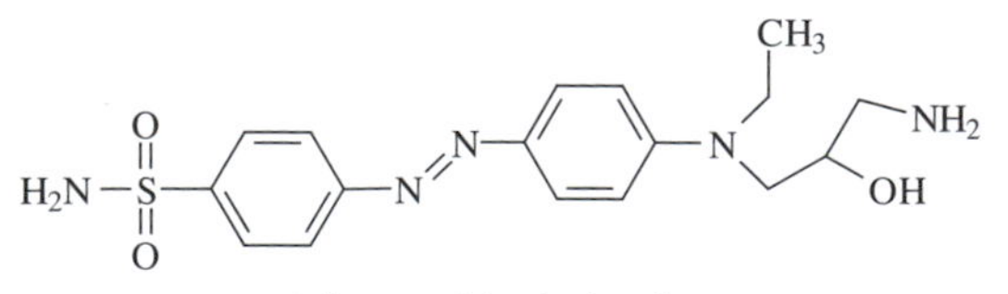

sulphonamide derivative

A few isolated clinical trials of the drug took place in Germany during the period 1933–1935, and Domagk wrote a number of reports about his discoveries; but the first serious clinical trial was carried out in 1936 at Queen Charlotte's Maternity Hospital in London. Here, Leonard Colebrook was trying to reduce still further the number of deaths from puerperal fever (which then stood at 2–3 per 1000 deliveries). Initially, he treated 38 women who were very seriously ill with streptococcal infections and saved the lives of all but three of them. Normally, at least 25% of these patients would have died. His *Lancet* paper generated considerable interest, as did a subsequent trial with a further 26 women, all of whom were saved.

Most of the subsequent development work on Prontosil Rubrum took place in France or Britain. The Rhone-Poulenc and Rousell companies made and supplied the drug, and in 1935, researchers at the Pasteur Institute in Paris showed that after administration, like other azo-sulfonamides, it was metabolised to sulfanilamide and that this was the active species. Since this was a widely known and readily available bulk chemical, no patent could be obtained, and any pharmaceutical company could produce and sell the drug. Various preparations were marketed and, in the absence of the stringent regulations that exist today, there was at least one major catastrophe. In the USA, Elixir of Sulphanilamide was sold as a solution of the drug in the solvent diethylene glycol, not realising that this solvent was a potent liver toxin. As a result, 76 patients died before the preparation was removed from the market; but there was one lasting benefit from this disaster and that was the enacting of a Food and Drug Act and, ultimately, the creation of the Food and Drug Administration. It was the existence of this very strict regulatory authority that allowed the USA to escape the catastrophe that occurred in Europe – with the drug thalidomide – some 20 years later.

Despite this setback, the pharmaceutical industry was encouraged to prepare literally hundreds of sulfonamides, of which May and Baker 693 (sulfapyridine) proved to be the most potent and broad spectrum. It also achieved star status once it was revealed that it had been used to save the life of Winston Churchill when he contracted pneumonia during a visit to North Africa in December 1943. Other sulfonamides that have been widely prescribed are sulfadiazine, sulfadimidine (especially for urinary tract infections and meningitis caused by meningococcal infections) and sulfamethoxazole. One problem with many sulfonamides is their relative water insolubility and their tendency to crystallise in the kidney tubules. They are also metabolised via acetylation of the aniline nitrogen, and these metabolites are both inactive and less soluble.

As for Gerhard Domagk, he received due recognition for his discovery with the award of the Nobel Prize for Physiology or Medicine in 1939, although, due to the exigencies of the Second World War, he was unable to

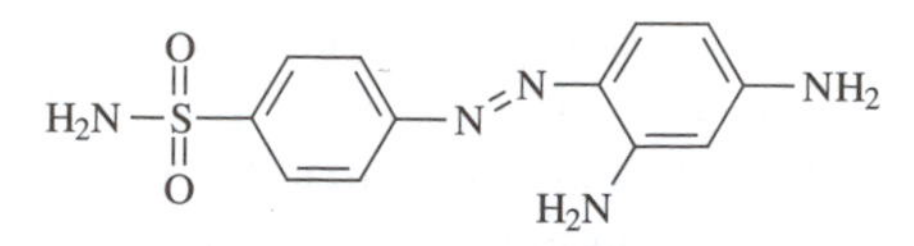

prontosil rubrum (Streptozon)

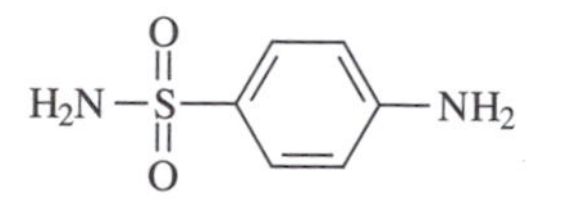

sulfanilamide

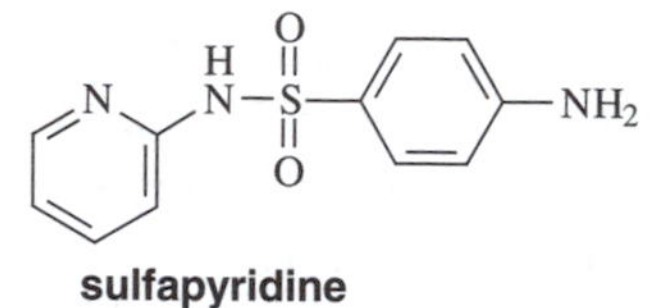

sulfapyridine

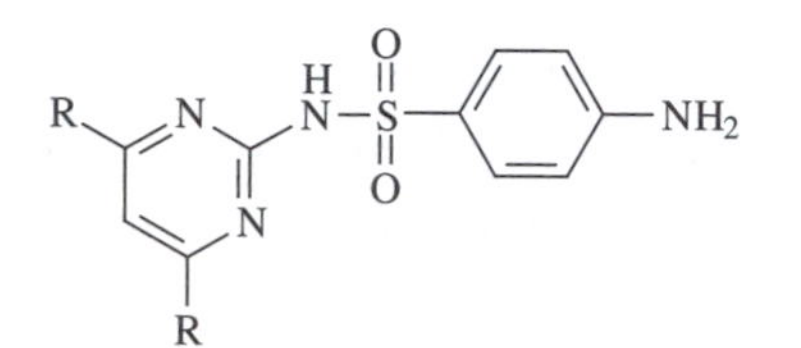

sulfadiazine (R = H)
sulfadimidine (R = CH_3)

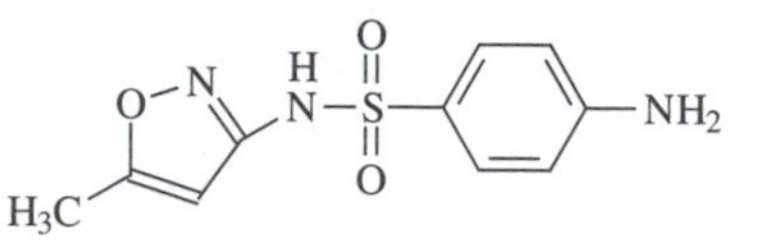

sulfamethoxazole

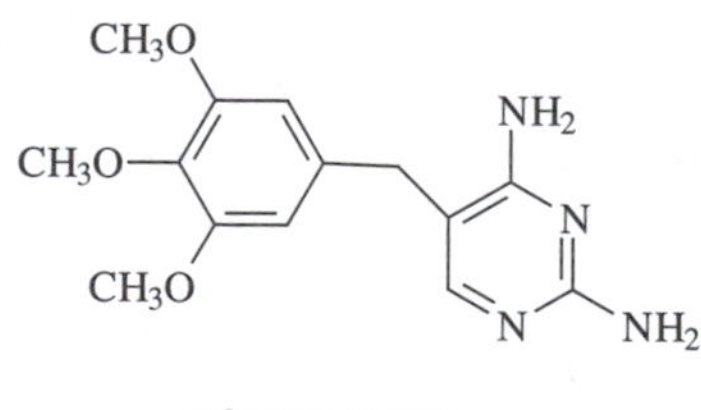

trimethoprim

collect his prize until 1947, by which time the prize money had reverted to the Nobel Foundation. But a greater personal satisfaction was undoubtedly attained when he was able to treat his daughter when she contracted septicaemia after pricking herself with a needle. In later life, he worked on drugs for the treatment of tuberculosis and was involved in the discovery of isoniazide, by far the best drug for the treatment of TB available in the 1950s. He was clearly a man who made major contributions to the chemotherapeutic armoury of the first half of the 20th century, but whose humility is revealed by a statement he made towards the end of his life:

> I am continuing to work in the field of chemotherapy, although I know that in all probability I will never be able to help as many people who will possibly be annihilated by a single atom bomb.

As for the mode of action of the sulfonamides, this was unravelled through a combination of serendipity and very clever detective work by Donald Woods at Oxford in 1940. He observed that the simple benzene derivative, 4-aminobenzoic acid, was an antagonist for sulfanilamide, that is, it allowed bacteria to grow even in the presence of the drug. This was ultimately shown to be due to the fact that bacteria have an absolute requirement for 4-aminobenzoic acid as an essential starting material for their production of dihydrofolate and tetrahydrofolate. These compounds are required for the production of DNA by all organisms, but whereas bacteria produce their own dihydrofolate, and thence tetrahydrofolate, mammals prepare these essential cofactors from dietary folic acid (present in fresh vegetables, eggs, liver and whole grain cereals). The sulfonamides act as inhibitors of one of the key bacterial enzymes of dihydrofolate production (dihydropteroate synthase), which catalyses the reaction of 4-aminobenzoate with 2-amino-4-hydroxy-6-hydroxymethyl-7,8-dihydropteridine pyrophosphate to produce dihydropteroate. In the presence of sulfonamides, the enzyme catalyses the synthesis of the sulfonamide analogue and the pathway to dihydrofolate is disrupted. Mammals do not possess this enzyme; thus, sulfonamides clearly have no effect on mammals and this makes these antibacterial drugs true 'magic bullets' in Ehrlich's terminology.

The unravelling of the whole biosynthetic pathway to dihydrofolate (shown in Fig. 2.1), mainly by the research group of George Hitchings and Gertrude Elion at the Wellcome Laboratories in Tuckahoe, New York, laid the foundations for their seminal work on inhibitors of nucleic acid (*i.e.*, DNA) biosynthesis. This was to lead to the discovery of a number of anticancer drugs and to another widely used antibacterial drug, trimethoprim. This inhibits another enzyme of the dihydrofolate pathway – dihydrofolate reductase – and trimethoprim is usually used in combination with the sulfonamide sulfamethoxazole under the tradename Septrin or Bactrim. This combination effectively inhibits two enzymes of the pathway, and it is usually more difficult for bacteria to become resistant to drugs when two separate inhibitory mechanisms are involved. However, despite these difficulties, many strains of bacteria have now become resistant to sulfamethoxazole, and trimethoprim is increasingly being prescribed by itself.

All the sulfonamides are bacteriostatic rather than bacteriocidal (they arrest growth rather than kill the bacteria), and while they made a major contribution to the quality of life in the 20th century, these contributions are dwarfed by those of the true bacteriocidal antibiotics like the penicillins and cephalosporins.

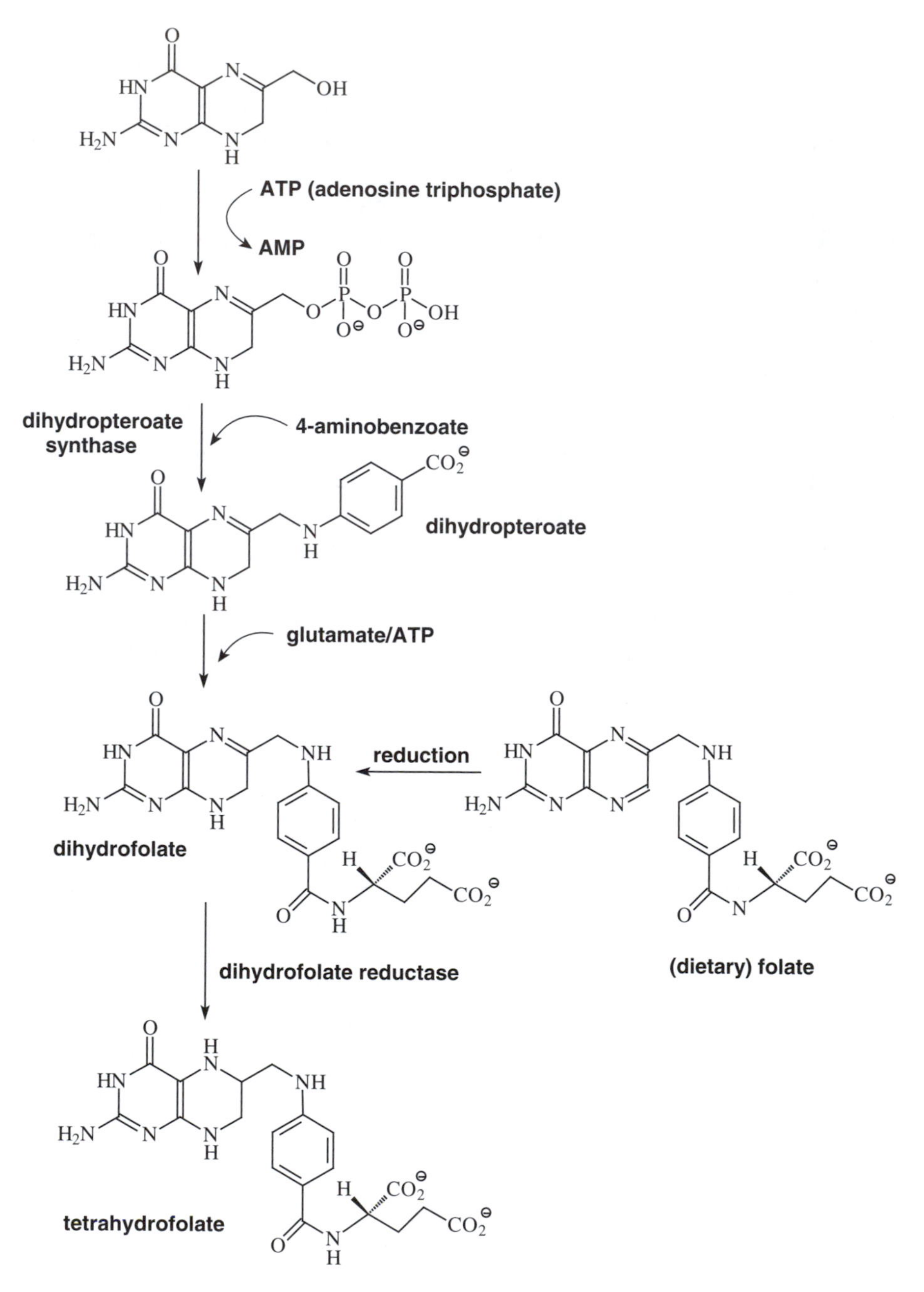

ATP (adenosine triphosphate)
AMP
dihydropteroate synthase
4-aminobenzoate
dihydropteroate
glutamate/ATP
reduction
dihydrofolate
(dietary) folate
dihydrofolate reductase
tetrahydrofolate

Figure 2.1

PENICILLINS: EARLY IDEAS OF ANTIBIOSIS

Pasteur was probably the first to observe that certain microorganisms could prevent the growth of other microorganisms, or '*la vie empêch la vie*'. In fact his work on anthrax with his colleague Joubert led them to make, in 1887, the following observation:

> Neutral or slightly alkaline urine is an excellent medium for the bacteria (anthrax)...if when the urine is inoculated with these bacteria and an aerobic organism, for example one of the 'common bacteria', is sown at the same time, the anthrax bacterium makes little or no growth and sooner or later dies out altogether. It is a remarkable thing that the same phenomenon is seen in the body of those animals most susceptible to anthrax... . These facts perhaps justify the highest hopes for therapeutics.

This phenomenon of one microorganism preventing the growth of another, or actually killing it, should come as no surprise, since the bacteria are amongst the oldest organisms in the biosphere. The various species have had to compete with one another for nutrients and space (their own ecological niche) for around 3.5 billion years. Bacteria do not possess an immune system like mammals, and of the many mechanisms that have evolved, the one of major interest to us is their use of chemical compounds that cause growth inhibition or death of a rival microorganism.

The French biologist Paul Vuillemin, in 1889, was the first to use the term *antibiosis* to describe this phenomenon whereby one organism destroys another in order to survive; he also used the term *antibiote* to describe the chemical substance involved. (It should be noted that the term *symbiosis* is used to describe the phenomenon whereby one organism cooperates with another to their mutual benefit.)

It was Selman Waksman, in 1941, who finally provided the definition of *antibiotic*, which is still accepted today. This stated that 'an antibiotic is a chemical substance produced by a microorganism that has the capacity to inhibit the growth and even destroy bacteria and other microorganisms.'

Various other observations of antibiosis were made prior to Fleming's seminal discoveries of 1928. In 1885, both Babes and Catani quite independently reported that certain bacteria could inhibit the growth of other species. The latter even claimed to have successfully treated a patient suffering from tuberculosis by causing him to inhale a crude mixture of bacterial species and gelatin. Two years later, the Swiss bacteriologist Garre showed that if *Bacillus fluorescens* was inoculated onto a culture plate containing *Staphylococcus pyogenes*, it would antagonise the growth of this bacterium. Of more potential clinical utility were the independent demonstrations by Freudenreich and Bouchard that *Pseudomonas pyocyanea* (now *P. aeruginosa*) would prevent the growth of typhoid bacilli and anthrax, respectively. An antibacterial principle, called pyocyanase, was isolated in 1899 from the same organism by Emmerich and Low and this was used therapeutically during the first few years of the 20th century. It was claimed to have good clinical activity against typhus, diphtheria and various infections caused by staphylococci; but its discoverers never revealed how they made their pyocyanase and eventually (and inexplicably) the commercial preparations ceased to exhibit antibacterial activity. More recently, it has been suggested that the antibacterial principle was one of the phenazine antibiotics, now called pyocyanine.

During the early years of the 20th century, other researchers claimed that various bacilli, including *Bacillus subtilis, B. mesentericus* and *B. megatherium*, showed useful clinical activity against tuberculosis; but much of this work was not reproducible. Of greater significance was the work of Lieske, and Gratia and Dath in the 1920s, who independently demonstrated that soil microorganisms of the family Actinomycetes could, in most instances, produce antibacterial substances. The culture plate of Gratia and Dath, showing zones of inhibition of staphylococcal growth in the area innoculated with a strain of actinomycetes (in the shape of an A), provided a graphic illustration of this phenomenon.

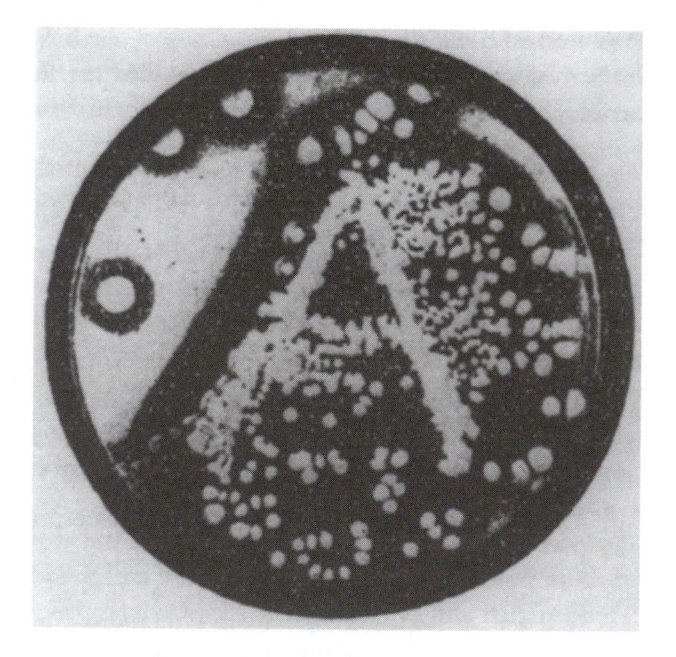

Demonstration by Gratia and Dath (1924 and 1934) of the lytic power of an actinomycetes. The gelatin contained a heavy emulsion of staphylococci. After 3 days, as shown, there was lysis of the staphylococci round the actinomycetes (planted in the shape of an A), and after 7 days the plate had cleared completely.

All of this work pre-dated Fleming's studies and it has often been said that Fleming received undue credit for what was, in essence, an extension of these earlier discoveries. However, this view ignores the most significant feature of his discovery. None of these other investigations gave rise to a major antibiotic, and although Fleming never produced therapeutically usable amounts of penicillins, his discoveries did open the door for others (Chain, Florey and their colleagues) to produce this major class of antibiotics.

PENICILLINS: FLEMING'S SERENDIPITOUS DISCOVERY

Penicillium moulds are most commonly encountered as greenish growths on the surface of contaminated jams and other preserves. The contaminating airborne spore germinates to produce thread-like structures (hyphae) that form branches with the appearance of a spoked wheel. Some of the hyphae undergo a transformation into a reproductive form with a thick stem that has the appearance of a pencil (hence the name penicillin), which branches to give a tree-like appearance. At the termini of these branches, spores develop, and when mature these are released into the air, and the life cycle of the mould is then complete. It is worth mentioning that the penicillium moulds are part of the large group of microorganisms that we usually call fungi.

Fleming was certainly not the first to study the biological properties of these moulds. John Burdon-Sanderson, a surgeon at St Mary's Hospital, Paddington, showed in 1870 that bacteria did not grow in the presence of contaminating *Penicillium* moulds. Lister experimented specifically with *Penicillium glaucum* and demonstrated that this mould prevented the growth of certain bacteria growing in samples of his own urine. He is also alleged to have treated one of his nurses, in 1882, with a mould extract after she suffered a septic finger following a pin-prick. Various other researchers experimented with *Penicillium* moulds, and the most success was obtained with *Penicillium brevicompactum*, whose antibacterial properties were subsequently shown to be due to the chemical constituent mycophenolic acid, which is presently of more interest for its antitumour activity and its immunosuppressant activity rather than as an antibiotic. But these were all 'minor skirmishes' with this family of moulds, and Fleming's work was much more important.

The story of the discovery of penicillins has been related on many occasions and in many forms, and the superb biography of Alexander Fleming by Gwyn MacFarlane gives by far the best account of his life and discoveries. It is, however, worth repeating the most salient features of the story, because it provides an excellent example of what Pasteur had earlier described as 'chance favours only the prepared mind' ['*le hasard ne favorise que les esprits preparés*'].

Alexander Fleming (Alec to his friends and family) was born in Ayrshire, Scotland in August 1881. The son of a farmer, he left school when he was 16 and became a shipping clerk in the City of London, a job that he seems to have tolerated rather than enjoyed. The boredom was partially alleviated when he joined the volunteer section of the London Scottish regiment of the army in 1900, along with his brothers John and Robert. This was at a time when the British army was heavily involved in the conflict with the Boers in South Africa, and fate could have decreed that he should participate in this conflict (with who knows what consequences); but he and his brothers were spared this eventuality. Another brother, Tom, was already a successful GP in London and persuaded Alec that he should become a medical student. This was not straightforward because Fleming had no qualification in Latin, and as a consequence, he had to attend night school for one year and studied not only Latin but also Scripture, Book-keeping and Shorthand amongst other more conventional subjects. He passed all his exams with excellent results, and entered St Mary's Hospital Medical School in October 1901 to start a medical career that would span 51 years in this hospital.

After his preclinical training, during which he studied pharmacology and organic chemistry, as well as the more obviously relevant physiology and anatomy, he progressed onto the wards of St. Mary's in 1904. In these early days of the Edwardian era, the situation in hospital wards was not very different from what had existed at the beginning of Victoria's reign some 85 years earlier. Most of the patients were suffering from potentially life-threatening complications of bacterial infections: meningitis, pneumonia, rheumatic fever, peritonitis and septicaemia. The average life expectancy (45 years) and childhood mortality (around 15% from birth to five years) had hardly changed at all during Victoria's reign. Fleming was not apparently tempted at this stage by the new discipline of bacteriology, but seems to have been more polarised towards surgery. Fate, however, played its part again since a position became available in 1906 in the Inoculation Department at St Mary's, and it was offered to Fleming, primarily to keep him in the hospital rifle shooting team. Fleming joined the team led by Almroth Wright, whose prime aim was to develop new vaccination regimes for the cure of bacterial infections. He was, for example, one of the pioneers of vaccination against typhoid and was responsible for persuading the British Army to carry out prophylactic vaccination against the disease at the commencement of World War I.

During the next eight years, Fleming learnt much about the techniques of vaccine preparation and of bacteriology, and even produced his own vaccine against the bacterium *Corynebacterium acnes*, which is usually involved in facial acne. His promising therapeutic results were described in a paper for the *Lancet* (1909), and his involvement with the clinical trials of Salvarsan (for the treatment of syphilis) has already been described in Chapter 1. This work provided an excellent training for his subsequent bacteriological research.

In October 1914, Almroth Wright and his team were sent to Boulogne to set up a research unit to study methods of treating infection of battlefield wounds. These usually contained fragments of metal, bone, clothing and soil. The latter was the most common cause of infection, and Fleming was personally responsible for identifying the presence of *Clostridium welchii* and *Clostridium tetani* in a high proportion of infected wounds. Both these organisms were common in the fields of Flanders due to the rich complement of horse manure, and caused the condition known as gas gangrene. Airborne staphylococci and streptococci were also prevalent. He also convincingly demonstrated that the available antiseptics, like carbolic acid, were not effective for these wounds. They rarely penetrated to the seat of the damage and even if they did, they were more damaging to the white blood cells (especially phagocytes, whose job is to engulf and destroy bacteria) than they were to the bacteria. These observations convinced him of the need for new types of more selective antibacterial agents, and the serious deficiency of the available antibacterial agents was further reinforced by his experiences of the effects of the Great Influenza Pandemic of 1918/1919. This will be discussed in detail in Chapter 3, but suffice it to state that well in excess of 20 million people died during the months of the pandemic, primarily as a result of post-influenza bacterial infection. At that time, it was believed that influenza was caused by a bacillus – Pfeiffer's bacillus or the influenza bacillus and later as *Haemophilus influenzae* – although subseqently it was shown to be caused by a virus. Certainly, the life-threatening pneuomonia could be caused by this bacillus, although more commonly by various types of staphylococcus, streptococcus and pneumococcus. Whatever the cause, Fleming and numerous other medical staff watched in horror as previously healthy young people, many of whom had just survived appalling wartime experiences, became ill with fever and within hours were dead from pneumonia.

These alarming deficiencies in therapy for bacterial infections undoubtedly influenced Fleming in his choice of research area when he returned to St Mary's in January 1919. It is unclear as to how he discovered that nasal mucus had antibacterial properties; but this line of investigation led to more important work with other body secretions and, in particular, with tears. The so-called 'tear antiseptic' was obtained from volunteers who had lemon juice instilled into their eyes, and proved to be reasonably potent. This same antibacterial agent was subsequently (in January 1922) shown to be present in egg white and this became the primary source of what was eventually christened lysozyme. In his first paper on lysozyme (in *Proceedings of the Royal Society*, 1922, B, vol. 93, 306), Fleming reported that about three-quarters of the airborne bacteria he had

Cartoon from Punch (1922) by J.H. Dowd depicting the collection of 'tear antiseptic'. In reality, the boys had lemon juice instilled into their eyes, and the resultant tears were collected for extraction of lysozyme

tested had failed to grow on cultures treated with mucus, saliva or tears. He suggested that this explained why these particular organisms were not pathogenic to humans. The obvious corollary of this, although it was not noted in the paper, was that other bacteria were pathogenic by virtue of the fact that lysozyme could not inhibit their growth. The preparations of lysozyme from egg white were more potent, and he was able to demonstrate that this could inhibit the growth of some pathogenic strains of streptococci, staphylococci, meningococci, typhoid, *etc*. A major advantage of this natural antibacterial substance was that, unlike the antiseptics, it did not damage white blood cells. Indeed, one of his collaborators, Frederick Ridley, showed that certain types of white blood cells actually produced lysozyme. However, the antibacterial potency of lysozyme would soon pale in comparison with that exhibited by the mould Fleming encountered in the late summer of 1928.

The precise origin of the *Penicillium* mould that contaminated one of Fleming's culture plates in August 1928 will never be known. It may have come in through the window of the laboratory, but more likely from one of the other rooms in the building; but whatever its origin, there is no question that it was a rare strain that was a highly efficient producer of penicillins. In addition, for the first week of August 1928, the temperature was unseasonably cold (ca 15 °C), which favoured growth of the mould. Then, the temperature soared to produce a heat wave, and this favoured growth of the staphylococci – except in the vicinity of the mould. The sheer serendipity of this contamination and of the ideal weather conditions was almost wasted when Fleming came close to destroying the evidence on his return from holiday. In his efforts to clean up the clutter left in the laboratory during his absence, he piled the used cultures into a tray containing the antiseptic lysol as a prelude to cleaning them for reuse. Had the key culture plate been submerged beneath the lysol, the evidence would have undoubtedly been destroyed. What he observed was described in his subsequent paper (*British Journal of Experimental Pathology,* 1929, vol. 10, 226): "around a large colony of a contaminating mould, the staphylococcal colonies became transparent and were obviously undergoing lysis".

His collaborators and others at St Mary's initially seemed unimpressed by this plate, since similar plates had been seen when lysozyme was employed; but what seems to have excited Fleming was the fact that the culture plate had been inoculated with a virulent strain of staphylococcus. This implied that the mould was producing an agent or agents that were effective against a potentially very pathogenic organism.

Fleming sub-cultured the mould, and through a series of experiments showed that it grew best at room temperature. He obtained a supply of antibacterial extract by allowing the mould to grow on the surface of a solution containing meat extract, and then filtering off the solids to leave crude 'mould juice'. This exhibited impressive potency against a large array of pathogenic bacteria, including virulent strains of staphylococcus, streptococcus, pneumococcus, meningococcus, gonococcus and the diphtheria bacillus. A simple name had to be found for 'mould juice', and Fleming coined the name *penicillin* to replace this term, fully aware of the fact that he was probably dealing with a mixture of antibacterial substances.

Alexander Fleming in his laboratory as St Mary's Hospital (From the photographic archives, St Mary's Hospital, Paddington)

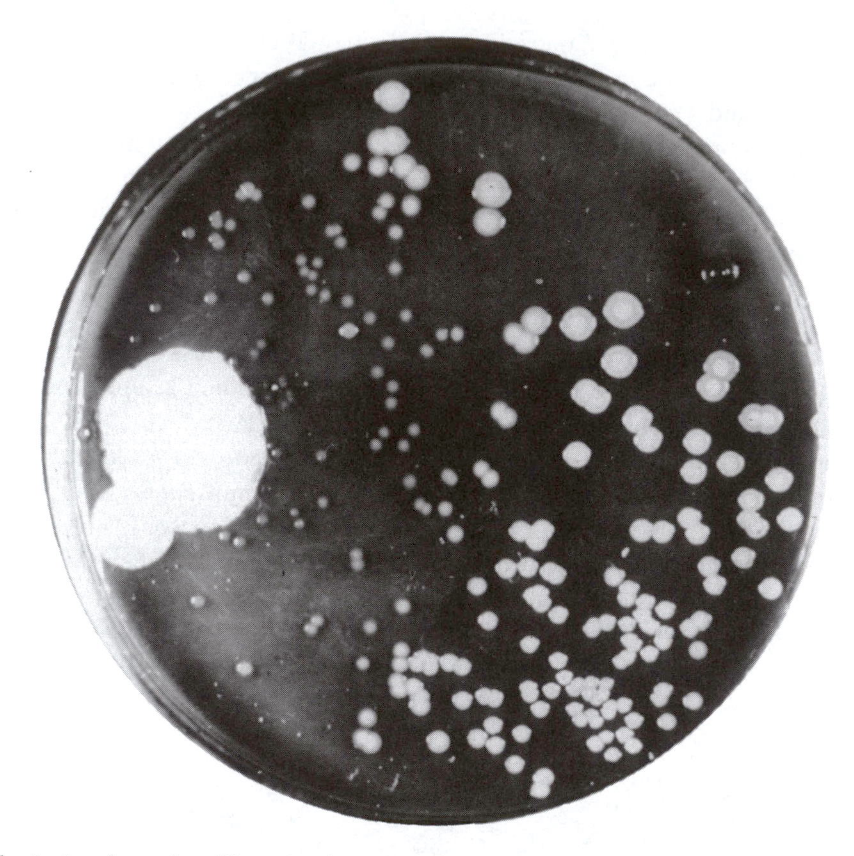

Fleming's culture plate (From the photographic archives, St Mary's Hospital, Paddington)

He initially thought that the mould was a potent strain of *Penicillium rubrum*, although it was subsequently shown to be *Penicillium notatum*. He screened a number of other *Penicillium* moulds from contaminated foodstuffs, old books, paintings, *etc*., but none except one provided by the hospital mycologist had the potency of the first mould. This confirmed his suspicion that he was dealing with a very unusual strain.

Meanwhile, his collaborators, Frederick Ridley and S. Craddock, had succeeded in improving the technology for the growth of the mould. They also managed to remove most of the water from the crude extract by evaporation under reduced pressure. This provided them with what Fleming called a 'sticky mass', which possessed most of the antibacterial activity originally present in the crude extract. Additionally, this product was soluble in alcohol, which meant that penicillin was unlikely to be a protein or enzyme, since these are always insoluble in this solvent. The chances of it being a relatively small chemical species were also more likely. On the negative side, the substance was relatively unstable at room temperature and was also sensitive to the pH of the growing medium. If this became even slightly alkaline (pH>7), the yield of penicillin was considerably reduced.

The antibacterial potency was also rapidly lost if penicillin was incubated with blood serum; hence, the chances of this new antibacterial agent having clinical utility seemed rather remote. In 1929, they did try a number of experiments on human subjects. One hospital patient, who was dangerously ill with septicaemia following an amputation, had his wound treated with penicillin, but with no apparent effect and the man subsequently died. They did claim one small success (although there is no formal written record of this) when they treated a case of pneumococcus-induced conjunctivitis with penicillin, and the infection was rapidly cured. But these were examples of penicillin being used as a topical antiseptic and not as a systemic antibiotic (that is, one that is carried through the bloodstream to sites of infection). If Fleming had tried injecting the penicillin into animals infected with bacteria (as the Oxford group did 12 years later), the antibiotic era might have commenced in 1929.

However, the apparent lack of clinical potential probably encouraged Fleming to exploit what he saw as the main use for his mould, and that was as a bacteriological tool for demonstrating the presence of Pfeiffer's bacillus, *i.e.*, *Haemophilus influenzae.* This was one of the few organisms that would quite happily grow in the presence of the mould, even when in competition with other pathogenic bacteria. Pure cultures of the organism could thus be obtained, and it is likely that Fleming had plans to prepare a specific vaccine against purified bacillus.

This aspect of his discovery was emphasised in his publication in the *British Journal of Experimental Pathology* (1929, vol. 10, 226), and also in a talk he gave to the Medical Research Club in February 1929. Fleming was, from most accounts, a poor public speaker and his talk elicited very little interest. His paper, although well written and containing full details of most of his experiments, seemed at that time to contain no particularly exciting revelations. As has already been mentioned, others had demonstrated the potential antibacterial properties of *Penicillium* moulds, and the idea of one microorganism inhibiting the growth of another was also well established. Fleming was also very modest in his claims for penicillin:

> Penicillin, in regard to infections with sensitive microbes, appears to have some advantages over the well-known chemical antiseptics. A good sample will inhibit staphylococci, *Streptococcus pyogenes* and pneumococcus in a dilution of 1 in 800. It is therefore a more powerful inhibitory agent than is carbolic acid and... it is non-irritant and non-toxic.

It was hardly an inspiration to the drug industry to develop penicillin as a major new antibacterial substance, although it is clear from Fleming's later writing that he genuinely believed he had discovered something of considerable novelty. For example, in his book entitled *Penicillin* (published in 1950 with contributions from many of the pioneers of penicillin production), he states:

> Penicillin was the first substance ever encountered...which destroyed bacteria without any apparent destructive action on the leukocytes. It was this observation more than any other which stimulated me in my first paper on penicillin to say "It may be an efficient antiseptic for application to, or injection into, areas infected with penicillin-sensitive microbes".

His evident frustration at being unable to make further progress is conveyed by his statement that:

> I had failed to advance further for the want of adequate chemical help. Raistrick (the Professor of Biochemistry at the London School of Hygiene and Tropical Medicine) and his associates had lacked bacteriological co-operation, so the problem of the effective concentration of penicillin remained unresolved.

Harold Raistrick had succeeded in growing Fleming's mould on a simple culture medium based upon glucose and some mineral salts, but had experienced severe difficulties in isolating the penicillin produced. These problems, both chemical and bacteriological, would not be solved for a further decade. Raistrick did help to identify the mould as *Penicillium notatum* (rather than *P. glaucum*) with the assistance of Charles Thom of the US Department of Agriculture (who was to play a major role in the scaleup of penicillin production).

In the meantime, Fleming maintained his special strain of *Penicillium notatum* in continuous culture, primarily as a bacteriological tool for his work with *Haemophillus influenzae*. Once again, fate was on the side of humanity, since it is unlikely that Fleming would have kept the mould if it had not possessed this useful property. The world might then have had to wait longer than the ten years that now elapsed.

CHAIN AND FLOREY AND THE PENICILLIN PRODUCTION LINE

Just like Fleming, Florey's pathway to the penicillins began with an investigation of lysozyme. Howard Florey was born in September 1898 in Adelaide, Australia, the son of an English immigrant who had made a fortune in shoe manufacture. After a very successful school career, he entered Adelaide Medical School in 1916 and, upon graduating, proceeded to study at Oxford as a Rhodes Scholar. He appears to have enjoyed the Oxbridge life and two years later, in 1924, took up a research fellowship in the Department of

Pathology in Cambridge. Here, he made several useful contributions in research on inflammation, gastric function and secretion of mucus; this latter area of research led him on to a major study of lysozyme structure and function that lasted until 1938. From Cambridge, Florey moved to the University of Sheffield as Professor of Pathology in 1932, but this proved to be a brief stay since he was offered the Chair of Pathology at the Sir William Dunn School of Pathology in Oxford three years later. This was to become the focus of penicillin research some four years later.

The work on lysozyme proceeded without interruption, and by 1937 they had enough pure material to allow Edward Abraham and Robert Robinson (Professor of Organic Chemistry in the Dyson Perrins Laboratory Oxford) to crystallize the enzyme for X-ray crystallographic studies. Additionally, this pure material allowed another new collaborator, Ernst Chain, a talented chemist and refugee from Hitler's Germany, to establish the fundamental chemistry of the antibacterial action of lysozyme. The actual importance of this natural substance in human defence systems was still unclear and Florey encouraged Chain to study other naturally occurring antibacterial products. Not surprisingly, one of these was *Penicillium notatum* and, since a sample of the mould (originally obtained from Fleming) was already in the Dunn School of Pathology, Chain and co-workers were able to attempt to repeat the published experiments of Fleming and Raistrick. That the Oxford team was eventually successful in obtaining clinically usable quantities of penicillin was due, in major part, to the experimental expertise of Norman Heatley, a biochemist whose brilliance at laboratory engineering tipped the balance between ultimate success and dismal failure. They were also aided in their endeavours by a generous grant of $5000 a year for five years from the Rockefeller Foundation, an award that must now rank as one of the most valuable investments ever made!

In 1940, Heatley invented the 'cylinder plate method' for assaying the strength of the penicillin. Short glass tubes were placed vertically in the culture containing staphylococci and then filled with the penicillin solution to be assayed. As the antibiotic diffused into the culture medium, there was a zone of growth inhibition (of the bacteria) whose diameter provided a measure of the strength of the penicillin solution. His second major innovation greatly improved the isolation of penicillin from the crude extract. In the earlier work of Raistrick and co-workers, the penicillin was separated from the water, after acidification, by extraction into either ether or amyl acetate. However, the acid instability of the penicillin led to severe losses during this extraction. Heatley devised a continuous flow apparatus in which the slightly acidic aqueous solution of penicillin flowed through a series of tubes, while ether (or amyl acetate) flowed in the opposite direction. The solvent extract was then passed down a second series of tubes while slightly alkaline water flowed in the opposite direction. Since penicillin contains a carboxylic acid group, the compound was extracted into the alkaline water as its salt. This could then be obtained as a brown powder by the process of freeze-drying. For this to work, the aqueous solution was frozen and a high vacuum was applied to the container, with resultant evaporation of the water at low temperature.

Using this new technology, the Oxford group managed to obtain about 100 mg of the brown powder (subsequently shown to have contained about 2% of penicillin and 95% of impurities!). After some preliminary toxicity tests on various animals, they carried out their first antibacterial tests on May 25, 1940. Eight mice were given a lethal dose of virulent streptococci and four were given two different doses of penicillin: two received one 10 mg injection, while the other two received five separate injections of 5 mg over a period of ten hours. The untreated mice were all dead within 24 hours, whilst three of the treated mice survived. The experiment was repeated on several further occasions using different doses of penicillin and other bacteria, and additional very positive results were obtained.

The significance of these results was not lost on Florey and his collaborators and they were quick to publish their results in the *Lancet* (August 24, 1940, vol. 2, 226) under the title 'Penicillin as a Chemotherapeutic Agent'. In this paper, they stated:

> The results are clear-cut and show that penicillin is active *in vivo* against three of the organisms (previously) inhibited *in vitro*. It would seem a reasonable hope that all organisms inhibited in high dilution *in vitro* will also be found to be dealt with *in vivo*.

The few tens of milligrams that had been used in these animal tests were but a fraction of what would be required for a clinical trial, and the team now

The 'penicillin girls' at the Dunn School of Pathology

turned their attention to scaling up the culture process. All kinds of shallow containers were tried, including biscuit tins, pie dishes and hospital bedpans. The latter proved to be ideal and the Staffordshire factory of James Macintyre and Co. Ltd were persuaded to produce several hundred (and later several thousand) ceramic culture vessels. These had a capacity of about one litre of culture fluid and were looked after by six 'penicillin girls'.

Large-scale production commenced on Christmas Day 1940 and rose eventually to an output of about 500 L of culture fluid per week, from which around 100–200,000 units of penicillin activity could be obtained. By the end of January 1941, enough penicillin had been accumulated to initiate a small-scale clinical trial. It is worth noting that these gargantuan efforts had been carried out while Britain struggled to survive following the evacuation from Dunkirk (in May 1940), the Battle of Britain (September) and the ever-present fear of invasion.

First, a terminally ill cancer patient in the Radcliffe Infirmary agreed to receive a 100 mg dose of penicillin (now about ten times more active per milligram than that used for the mouse trials). Florey and others, watching by her bedside, were distressed to see that she suffered an initial attack of pyresis – high temperature and shaking – although she was otherwise unaffected. Reasoning that this could be due to the presence of impurities, Edward Abraham was given the task of purifying the penicillin by the new technique of alumina chromatography. Finally, they had penicillin that was deemed pure enough for administration to a suitable patient.

They chose an Oxford policeman, Albert Alexander, who was desperately ill with advanced streptococcal and staphylococcal infections following a simple sore on his mouth (a further reminder of how precarious life was before the advent of antibiotics!). He had already received extensive treatment with sulfonamides, but to no avail. On February 12, he was given an intravenous injection of 200 mg of penicillin followed by further injections of 100 mg every three hours. Within 24 hours, he showed a dramatic improvement and within five days he was clearly well on the road to recovery. Sadly, the supplies of penicillin ran out and despite extracting his urine to recover some of the drug for recycling, his treatment could not be continued and he relapsed and died of septicaemia on March 15 – a timely indication that eradication of all bacteria must be achieved if a cure is to be obtained.

A further five, dangerously ill patients were treated between February and June and all but one of these were cured. Their other failure was a four-year-old boy who was seriously ill with infected measles spots. He showed a dramatic recovery with penicillin treatment, and all the staphylococci were eliminated; unfortunately, he had suffered from neurological damage during the viral infection, and this was the ultimate cause of death rather than the bacterial infection.

Despite these two deaths, the success rate with the penicillin was quite unprecedented, and the Oxford team wrote a second paper for the *Lancet* in August (vol. 2, 177). In this, they provided details of the production of penicillin as well as the results of all their animal and clinical trials. They also noted that they had observed the development of resistance to penicillin by certain bacteria, and this was to become a major problem with all antibiotics. The paper concludes with a rather modest statement about the clinical possibilities of the new drug:

> Enough evidence, we consider, has now been assembled to show that penicillin is a new and effective type of chemotherapeutic agent, and possesses some properties unknown in any antibacterial substance hitherto described.

Thirteen years, almost to the day, after Fleming's initial discovery, they had unknowingly announced the dawn of the antibiotic era.

It was quite clear that the scale of production in the William Dunn School could barely provide enough material for further clinical trials; hence, Florey tried to elicit the help of drug companies. ICI, Burroughs Wellcome, Boots and a small London company, Kemball-Bishop, all showed interest; but due to the exigencies of war-torn Britain, none had sufficient funds for speculative research. They were probably also worried that a chemical synthesis of penicillin would be devised once its structure had been elucidated; thus, money spent on culture technology would be wasted. They were also probably worried about the patent situation. Florey had mentioned this issue to the Medical Research Council, which had provided a modest amount of research money. Apparently, the senior MRC officials were vehemently opposed to patenting on the grounds that it was unethical for medical researchers to benefit from their discoveries. The American pharmaceutical companies suffered no such qualms of conscience, and in due course the British companies had to pay royalties to their US counterparts before they could produce penicillins.

In the end, ICI did produce some penicillin in a pilot plant at their dyestuffs division. In addition, Kemball-Bishop produced 150–200 gallon batches of culture extract and these were shipped to Oxford in milk churns for further processing.

It was perhaps inevitable that the Oxford team would turn to their colleagues in the USA for assistance, since America was rich and still at peace. Florey and Heatley travelled to the USA in July 1941 with samples of *Penicillium notatutm*. Almost at once, they received very constructive help from Charles Thom of the Department of Agriculture, the scientist who had helped with the identification of the mould some ten years previously. He passed them on to the USDA's Northern Regional Research Laboratory in Peoria, Illinois, which had considerable expertise with fermentation technology. They quickly established that the mould could be grown more effectively on corn steep liquor (a by-product of corn starch manufacture), and were thus able to raise the level of penicillin from the 1.8 mg L^{-1}, obtained with the Oxford culture medium, to 14 mg L^{-1} on corn steep medium. Of much greater significance was the discovery that the chemical structure of the major antibiotic produced had changed. Careful purification of what had always been termed 'penicillin' revealed that a number of different chemical species were present in the culture extract. The Oxford culture medium provided mainly what became known as penicillin F (subsequently shown to have a hex-3-enoyl side-chain), with small amounts of the more stable (and hence more clinically effective) penicillin G (with a benzyl side-chain). The corn steep medium mainly provided penicillin G. This selectivity could be further enhanced if a chemical precursor (phenylacetic acid) for the side-chain of penicillin G was added to the culture medium. Finally, and most imaginatively, they encouraged people in Peoria to submit samples of mouldy food for examination, and this resulted in the discovery of a strain of *Penicillium chrysogenum* (code name NRRL 1951) from a mouldy canteloupe in Peoria market. With the right growth conditions, this proved to be an even better source of penicillin G and (together with a few mutant strains) remains the mould of choice for the production of penicillin G – yet another example of how serendipity favoured the evolution of the penicillins.

Florey and Heatley also visited various US pharmaceutical companies and Merck, Pfizer, Squibb and Lederle all made plans to become involved in the project. This involvement became a national priority following the Japanese attack on Pearl Harbour in December 1941. Massive amounts of money and manpower were committed to a project that could save the lives of American servicemen and women. One major innovation was the development of deep tank fermentation technologies (rather like those used for brewing beer), and this required efficient mixing and aeration of the culture medium whilst maintaining an aseptic (*i.e.*, bacteria-free) environment. An idea of the success of this innovation can be gained from the fact that by the end of the war, the US company Pfizer (the largest producer) was producing nearly 100,000 million units of penicillin every month, with enough overall to treat about a quarter of a million patients.

The American public had been given a foretaste of this 'magic bullet' prior to its arrival on the battlefield. On November 28, 1942, a devastating fire at the Coconut Grove night club in Boston left around 500 people dead and many hundreds with serious injuries. Despite strict wartime secrecy, news leaked out that employees at Merck in Rahway, New Jersey, had worked around the clock to prepare enough of 'an unnamed miracle drug' in order to provide antibacterial treatment for these burn victims.

In Britain, production lagged very much behind that of the US, and for most of the war, all the penicillin came from the Oxford Group. They treated a further 187 cases in 1942 with almost total success; but it was the belated involvement of Fleming that finally encouraged the government to take an interest in penicillin. In August 1942, a family friend of the Flemings became seriously ill with meningitis, and Fleming asked Florey for samples of penicillin and instruction on its use. These were freely given and the treatment was successful:

> Florey was good enough to give me his whole stock of penicillin to try on this, the first case of meningitis to be treated. After a few days' treatment with intramuscular and intrathecal (*i.e.*, directly into the blood vessels of the brain) injections the patient was out of danger and he made an uneventful recovery.

The fortunate outcome of what had been a dangerous experiment obviously made a huge impression on Fleming, because he immediately contacted friends in the government and a Penicillin Committee was soon established. The pharmaceutical industry was finally galavanised into action, and pencillin production began in earnest.

As stories began to emerge about the efficacy of penicillin, the press printed stories about this new 'miracle cure', and about its inventors. Much of the mythology about Fleming arises out of misinformation printed at this time. More seriously, a certain degree of ill-feeling between the London and Oxford groups arose from ill-conceived comments about the relative importance of their respective discoveries. Several letters to the *Times* make interesting reading. Sir Almoth Wright, Professor of Bacteriology at St. Mary's Paddington when Fleming was making his seminal discovery, wrote on August 31, 1942 suggesting that Fleming should receive the laurel wreath "for he is the discoverer of penicillin and was the author of the original suggestion that this substance might prove to have important applications in medicine". Sir Robert Robinson responded on September 1 that "a bouquet at least, and a handsome one, should be presented to Professor Florey". This was all rather unfortunate and the controversy was resolved in 1945 when the Nobel Prize for Physiology or Medicine was awarded jointly to Fleming, Florey and Chain, a fitting reward for what had been one of the outstanding achievments of the 20th century.

SEMI-SYNTHETIC PENICILLINS AND THE GROWING PROBLEM OF RESISTANCE

The chemical structure of the penicillins remained in doubt throughout the war years despite considerable efforts and much speculation by a group led by Karl Folkers at Merck, and the Oxford group comprising Robert Robinson, Chain, Abraham and others. One of the major problems was that the initial microanalysis had not revealed the presence of a sulfur atom, and it was not until penicillamine was revealed in 1943 as the major acid degradation products of all the known penicillins that realistic structures could be proposed. The Oxford group (especially Robert Robinson) favoured an oxazolone-thiazolidine structure, while the Merck group preferred an alternative structure that contained a beta-lactam ring – the Oxford group also considered this as an alternative to their preferred structure. This second structure required a leap of faith because the beta-lactam ring had never been observed before in a natural product.

The structure was finally resolved in 1945 by an X-ray crystallographic study carried out by Dorothy Hodgkin on penicillin F at Oxford. This confirmed that the penicillin nucleus comprised a five-membered ring containing a sulfur atom, and this was attached to a four-membered beta-lactam ring. A variety of acyl side-chains were attached to the amino group. Thus, penicillin F had a hex-3-enoyl side-chain and penicillin G had a phenylacetyl side-chain. Penicillin G was thus also known as benzylpenicillin and this was the antibiotic that saved countless lives during the closing years of the war. It is interesting to note that, for a long time after the X-ray structure proof,

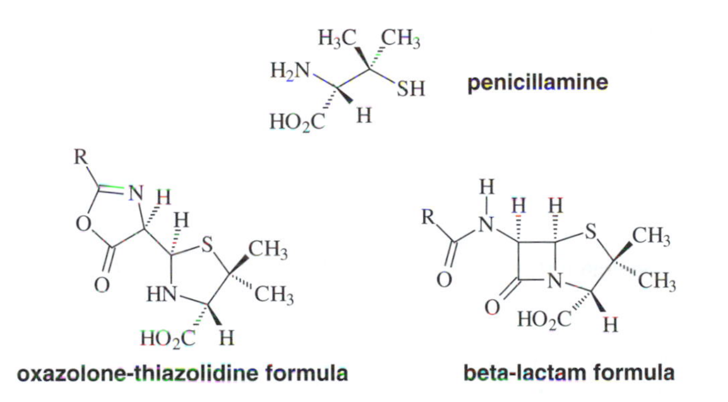

Robert Robinson refused to accept the beta-lactam ring, suggesting "that is the result when penicillin is in the solid form. You simply do not know what it is when the penicillin is in solution."

The discovery that addition of side-chain precursors to the culture medium favoured production of the penicillin that had incorporated this side-chain allowed the production of a whole range of novel penicillins. Although around 100 such structures were prepared, only one had more desirable properties than penicillin G, and this was christened penicillin V or phenoxymethylpenicillin. Although this compound was not as potent as penicillin G, it was much more stable to acid (in the stomach), and so could be administered by mouth rather than requiring injection like penicillin G. This was a major advantage for patients, and penicillin V rapidly became the antibiotic of choice for GPs.

But just as the elimination of bacteria-induced infections seemed a realistic possibility, the bacteria 'retaliated' against these new drugs. We should not be surprised by this since the bacteria have been co-evolving with one another and with other microorganisms (like moulds) for more than three billion years. They can reproduce every half an hour or so and can thus evolve at a much faster rate than humans. The first resistance mechanism to be observed involved the appearance of bacterial strains that produced penicillinase enzymes. These destroyed the beta-lactam ring of the penicillins, and since this is an absolutely vital part of their structure for effective antibiotic action, the drugs became ineffective.

Resistance by virtue of the production of penicillinases had been observed as early as 1940 by Florey and Abraham, and by 1946, around 15% of the

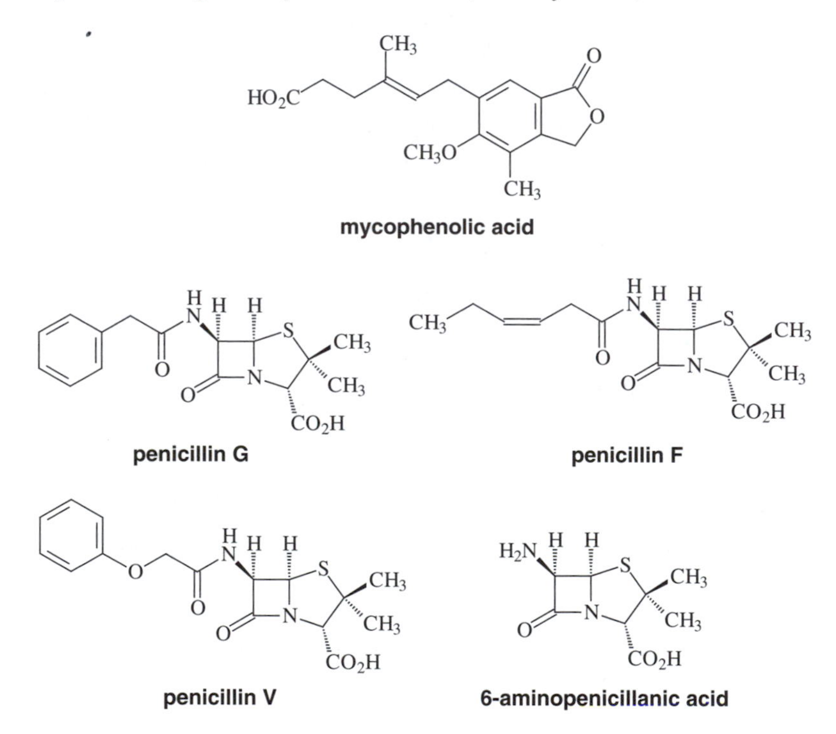

strains of *Staphylococcus aureus* isolated in London hospitals were resistant to penicillin G. By 1947 this figure approached 40%, and by 1948 it was 60%. Today, penicillin G is essentially useless against all strains of this bacterium. An increasingly desperate race then commenced to discover new penicillins that were able to destroy the rapidly evolving resistant strains of bacteria.

The first major breakthrough was achieved in 1958, when a team at Beecham Research at Brockham Park near Dorking discovered that the basic nucleus of the penicillin molecule (called 6-aminopenicillanic acid or 6-APA) could be isolated from the culture medium if the growth conditions were adjusted. Since 6-APA was lacking the side-chain of penicillin G or V, chemical synthesis could now be used to prepare virtually any penicillin structure that could be imagined. They also developed various enzymatic methods using aminoacylases for cleavage of the natural side-chain to produce 6-APA.

At around the same time, John Sheehan at the Massachusetts Institute of Technology prepared 6-APA and penicillin V by purely chemical means in 1957. Although his synthesis was relatively concise and his initial yield of 1% was increased to 60%, he was unable to prepare commercially usable amounts of these compounds. His synthesis was notable, however, for its use of dicyclohexylcarbodiimide (DCC) as the peptide coupling agent that was used to create the beta-lactam ring. DCC subsequently became the standard reagent for peptide coupling, and together with modern variants, is still used for this and other reactions where mild and neutral dehydration is required.

Due to the near contemporaneous discoveries by Beecham and Sheehan, a long legal wrangle ensued over who had the prior claim to have made 6-APA. Sheehan's US patent application was filed in March 1957, whilst the Beecham group filed a British patent application in August 1957. There was no doubt that the Beecham product was purer than the Sheehan product, and the British lawyers made the suggestion that the Sheehan patent had not provided a clear indication of how he had prepared 6-APA. The legal wrangling continued until February 1979, when the US Board of Patent Interferences finally awarded Sheehan the prior claim.

Later, chemical methods and more efficient enzyme-mediated methods were devised that could selectively remove the side-chain of penicillin G to produce 6-APA. These two commodities, 6-APA and penicillin G, thus became the high tonnage products of the fermentation process. Modern fermenters now range in capacity from 100,000 to 200,000 L and can provide 40 g L^{-1} of penicillin G (compared with less than 20 mg L^{-1} in 1941).

The novel penicillins that were produced in the 1960s were called *semi-synthetic penicillins*, since they were obtained by a combination of fermentation technology and chemical synthesis. Throughout the 1960s and 1970s, literally hundreds of semi-synthetic penicillins were prepared. Of these, ampicillin and amoxycillin were amongst the most successful since they combined broad-spectrum antibacterial activity against Gram-positive and Gram-negative bacteria with fair oral availability. Methicillin, oxacillin and cloxacillin, in contrast, had only weak antibacterial activity but retained their potency even in the presence of resistant (penicillinase-producing) strains of *Staphylococcus aureus*, and the last two also had good oral availability. It was subsequently shown that their relatively bulky side-chains prevented the binding of the penicillinases to the drugs. In recent years, it is the appearance of strains of methicillin-resistant *Staphylococcus aureus* (MRSA), especially in hospital environments, that has caused so much alarm. Other innovations involved the design of penicillin prodrugs like pivampicillin and talampicillin. They overcame the fact that only about 50% of ampicillin was absorbed from the gut, and these prodrugs passed through the stomach and

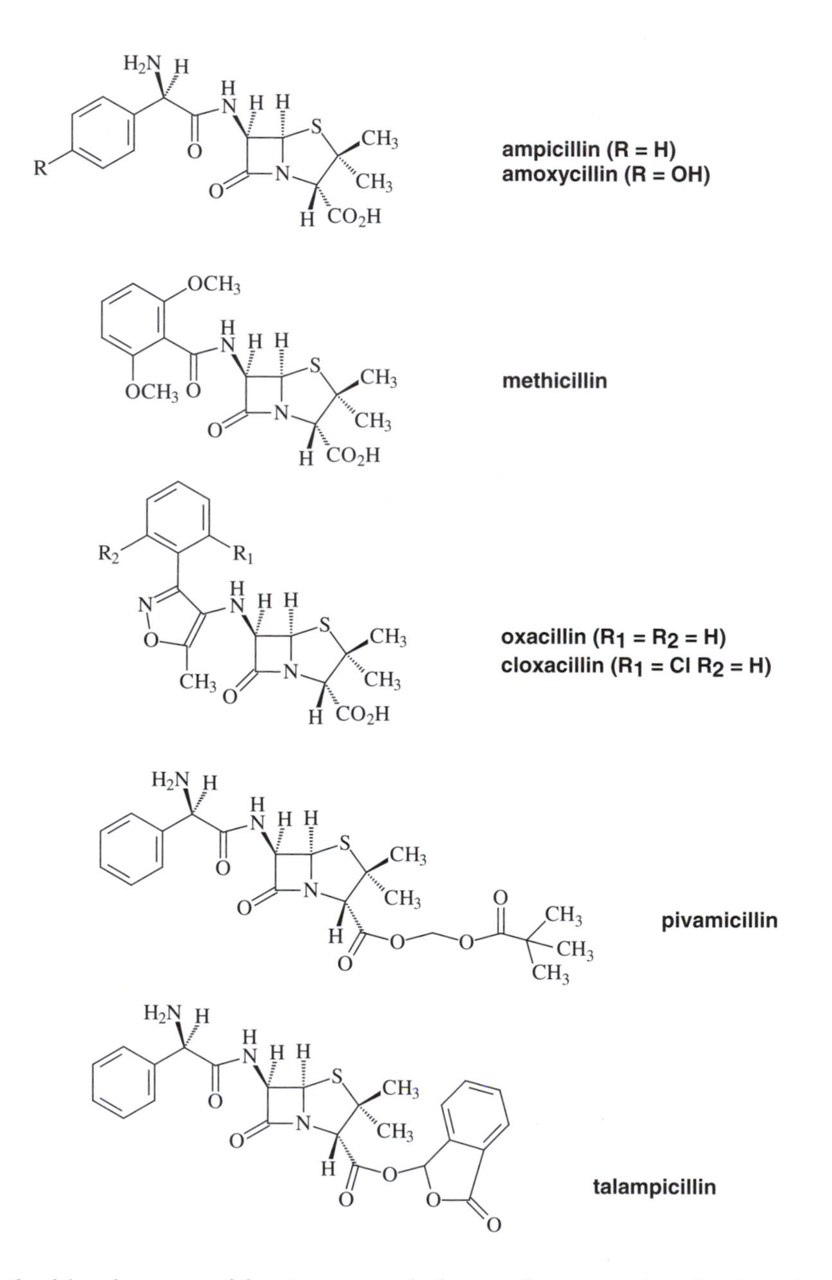

into the bloodstream without structural change, but were then broken down by enzymes in the bloodstream to produce the potent antibiotic ampicillin.

Despite the advent of all these excellent analogues of penicillin, it was evident, even 30 years ago, that the penicillins were not going to be the ultimate antibiotics. They were generally only active against Gram-positive organisms, and thus ineffective against typhoid or salmonella, *etc.*; resistance developed rapidly; and in a small percentage of the population, potentially life-threatening allergic reactions could be induced by the administration of penicillins. The other major class of beta-lactam antibiotics, the *cephalosporins*, are superior in all these respects to the penicillins, and their discovery was almost as serendipitous as that involved with the penicillins.

THE MOULD THAT GREW ON SEWAGE: EVOLUTION OF THE CEPHALOSPORINS

Guiseppe Brotzu, director of the Istituto d'Igiene in Cagliari, Sardinia, had read about the momentous discoveries made in London and Oxford, and was attracted to a peculiar mould that grew near the main sewage outfall in Cagliari. In 1945, he identified this as a strain of *Cephalosporium acremonium* and showed that a crude extract of the mould would inhibit the growth of typhoid bacilli and other bacteria in culture. He sub-cultured the mould and eventually obtained a strain that had activity against both Gram-positive and Gram-negative organisms, and was even bold enough to attempt a preliminary clinical evaluation of various extracts. When applied to boils and other surface infections, there were positive results, and he also injected his extracts into patients suffering from typhoid and paratyphoid infections. Again, there were some encouraging responses. He published his results (in Italian) as part of a report from his institution in 1948, and this came to the attention of a local British health officer who communicated news of the discoveries to Florey in Oxford. Brotzu was then encouraged to send a sample of his mould to the Dunn School of Pathology.

The isolation and structure elucidation of the most important constituent, cephalosporin C, took a further 13 years. In the early stages, the combined efforts of Norman Heatley, Edward Abraham, H.S. Burton and C. Newton led to the isolation of what proved to be another penicillin, christened penicillin N. Although this had reasonably good activity against Gram-positive and Gram-negative organisms, it was very unstable, and offered no advantages over the existing penicillins. Further purification of the crude extract revealed the presence of a novel substance, which also had broad-spectrum activity, albeit at a low level. It did have the advantage that it was less toxic than penicillin G and was also resistant to penicillinases. Abraham and Newton worked out the structure of this new antibacterial substance and proved that it possessed the same type of beta-lactam ring as the penicillins, but this was now fused to a six-membered ring containing a sulfur atom rather than the five-membered ring of the penicillins. Interestingly, this was one of the first natural product structures where proton NMR studies were of pivotal importance. Abraham and Newton noted in their *Biochemistry Journal* paper (1961, vol. 79, 377–393):

> In the spectrum of benzylpenicillin…a large peak at 7.9 ppm (relative to tetramethylsilane at 10.0 ppm) was due to the gem-dimethyl group. The spectrum of cephalosporin C showed no such peak. A peak of the intensity required (for one methyl group) was present at 7.4 ppm…a peak due to the (other) single methyl was in a position expected for the methyl of a CH_3.CO.O group.

It is interesting to note that this spectrum was obtained for them by (Sir) Rex Richards.

The novelty of this structure coupled with its interesting spectrum of antibacterial activity were clearly of potential commercial interest, and this time there were no qualms about patenting the discoveries. The strain of *Cephalosporium acremonium* was made the property of the National Research Development Corporation (NRDC) in Britain, and over the course of the next 15 years or so, this body received (on behalf of the British government) a fortune in royalties for the development of the cephalosporin antibiotics.

Considerable work went into modification of the strain (through the use of X-rays or UV light to induce mutations) and improvements were made in the technology of isolation and purification. However, the mould was much more difficult to handle and would not accept alternative side-chain precursors as did *Penicillium chrysogenum*; hence, it was impossible to prepare semi-synthetic cephalosporins by this route. In addition, the discovery of methicillin with its resistance to penicillinases negated the advantages of cephalosporin C, and for a while it appeared that this new class of antibiotics was doomed.

This situation changed when an efficient chemical method was developed at Eli Lilly in Indianapolis, which would remove the side-chain of cephalosporin C to produce a bare cephalosporin nucleus, the so-called 7-aminocephalosporanic acid, from which a whole range of semi-synthetic cephalosporins could be produced. Of these, cephaloridine and cephalothin were the most widely used. Of even greater significance was the invention of some very clever chemistry, also by the Lilly group, which involved the conversion of the penicillin nucleus into the cephalosporin nucleus in a short sequence of chemical transformations (Fig. 2.2). This meant that the very cheap penicillin G could be converted into what became the best-selling cephalopsorin cephalexin (Keflex). This had excellent oral bioavailability, was very stable to metabolism, and resisted the actions of penicillinases. The other semi-synthetic cephalosporin that came to be widely used was cefaclor

(Distaclor), but literally hundreds of other exotic structures have been prepared over the years. Primarily, these have to be injected and are usually reserved for hospital use, although cephalexin and cefaclor are both orally active and widely prescribed. Various more complex cephalosporins, like cefotaxime and ceftazidime, are especially valued for their activity against

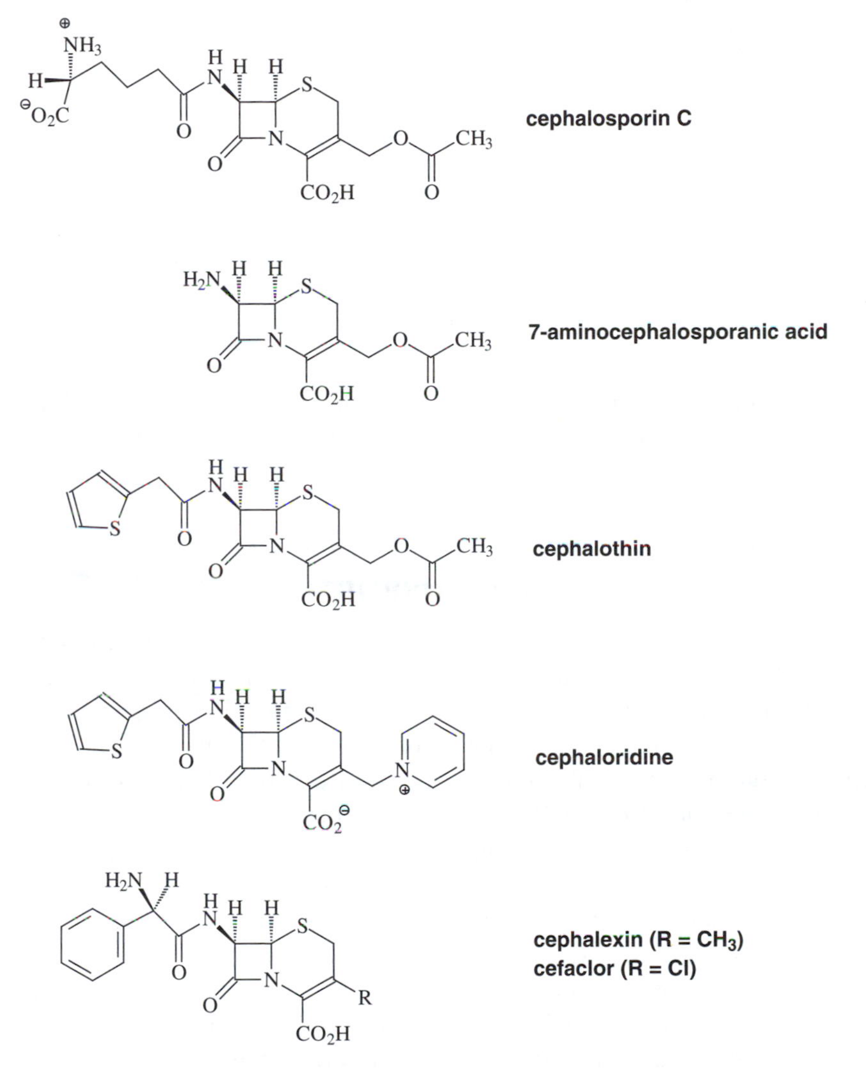

Gram-negative bacteria, and these two have spectacular activity against pseudomonads, which are particularly dangerous for patients with serious burn injuries.

The bacteria have responded to the advent of these drugs by producing cephalosporinases (which destroy the cephalosporins); but because the drugs

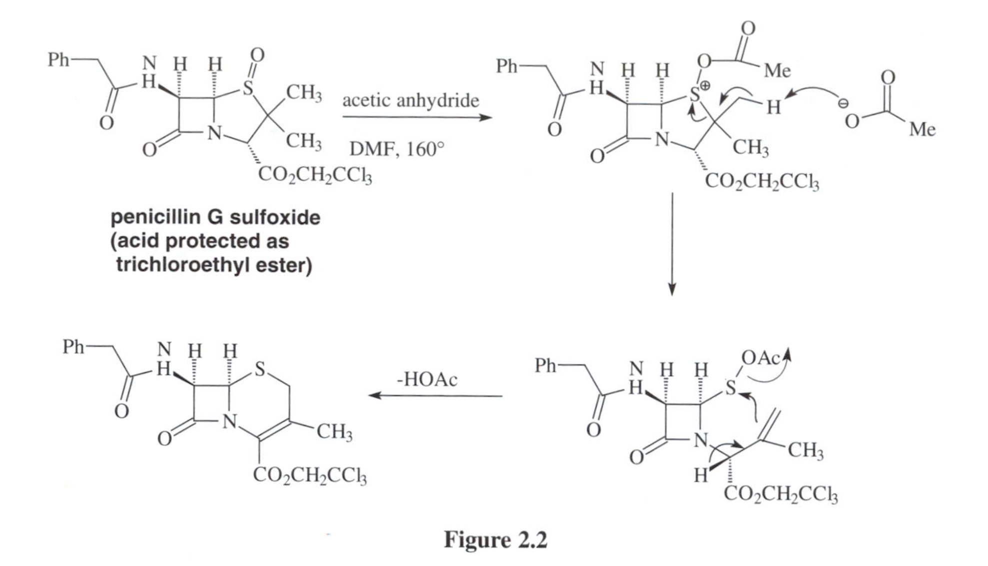

Figure 2.2

have not been as widely disseminated in the population as the penicillins, the bacteria have not encountered them as frequently, and the problem of resistance is not so grave.

THE MODE OF ACTION OF PENICILLINS AND CEPHALOSPORINS

The beta-lactam antibiotics all function by disrupting the formation of the cell walls of bacteria. As mentioned earlier, the bacterial wall of both Gram-positive and Gram-negative organisms comprises a glycopeptide matrix with other associated molecules like lipids or polyprenols. The basic unit of the glycopeptide is a molecule containing two different carbohydrates (usually *N*-acetylmuramic acid and *N*-acetylglucosamine) joined together to make a disaccharide, with a chain of amino acids attached to the *N*-acetylmuramic acid portion. In most bacteria, the polypeptide chain is five units long with a branching point about halfway along its length to which another chain of amino acids is attached (see Fig. 2.3). The main polypeptide strand almost invariably terminates with the same two amino acids – D-alanine attached to another D-alanine, and this is crucial for the mode of action of the beta-lactam antibiotics and also to several other types of antibiotics.

When a bacterium divides, it requires new cell wall material, and the first stage in the production of this material involves movement of the disaccharide-polypeptide moiety to the outside of the bacterium. At this stage, the carbohydrate portion is attached to a lipid that anchors this basic building block to the bacterial cell membrane. An enzyme (a transglycosylase) now

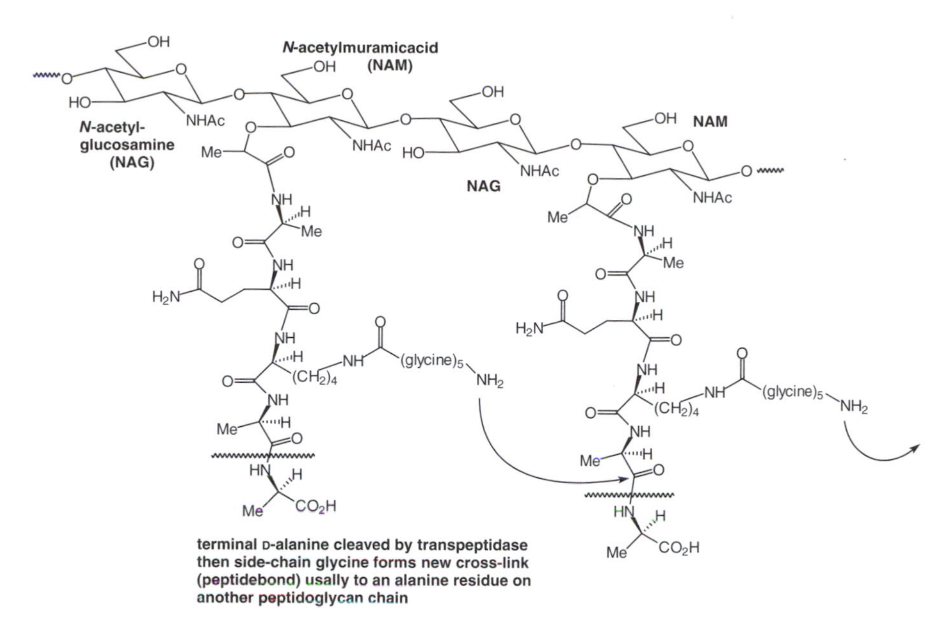

Figure 2.3

facilitates the joining together of the disaccharide portions to produce a polysaccharide chain from which the pendant polypeptides are suspended. Another enzyme, a transpeptidase, now catalyses the removal of the terminal D-alanine residues with concomitant reaction between the terminal side-chain amino acids and the residual D-alanines of the main polypeptide chains. In this way, a three-dimensional, cross-linked structure is produced, and this is the new bacterial cell wall material.

The three-dimensional structure of the beta-lactam portion of both penicillins and cephalosporins is sufficiently similar to that of D-alanine–D-alanine (see Fig. 2.4), that the transpeptidase enzyme acts upon the drugs instead of the bacterial polypeptide chain. The enzyme becomes covalently attached to the antibiotic and is then unable to carry out its normal functions such that new cell wall material cannot be produced and the dividing bacterium cannot survive. Unfortunately, the penicillinases and cephalosporinases (now known collectively as beta-lactamases) that have evolved to meet the threat posed by the antibiotics act upon the drugs to destroy their beta-lactam rings so that they can no longer inactivate the transpeptidase enzymes.

The pharmaceutical industry has responded to this problem by searching for other naturally occurring compounds that can destroy the beta-lactamases. Although a number of such compounds have been identified, only one of them has been a major clinical success. In 1975, Beecham isolated a compound that they christened clavulanic acid from the soil microorganism

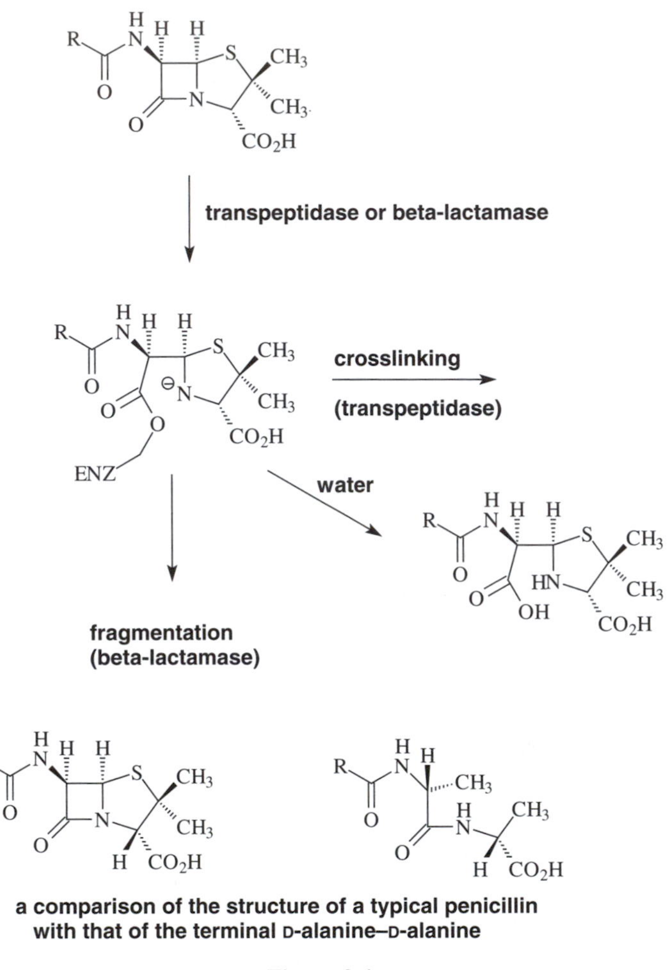

a comparison of the structure of a typical penicillin with that of the terminal D-alanine–D-alanine

Figure 2.4

Streptomyces clavuligerus. This was essentially devoid of antibacterial activity but was a very effective inhibitor of beta-lactamases produced by a wide variety of bacteria. Some idea of the excitement these results caused can be seen from Beecham's data for the minimum inhibitory dose of ampicillin needed for a typical resistant staphylococcus without clavulanic acid (500 μg ml^{-1}) and with clavulanic acid (less than 0.4 μg ml^{-1}). Five years later, they were able to demonstrate that the combination of the broad-spectrum penicillin – amoxycillin – with clavulanic acid was very effective in clinical use. This combination was marketed as Augmentin and soon became one of the best-selling drugs in the world market.

Its efficacy derives from the fact that the clavulanic acid acts as a sacrificial substrate for the beta-lactamases and thus prevents them from destroying the amoxycillin (see Fig, 2.5). This well-established penicillin can then function as an inhibitor of cell wall assembly. Unfortunately, it is such an effective antibacterial combination that it destroys many of the useful bacteria that reside in the human gastrointestinal tract as well as the pathogenic organisms. This is true of many of the new, highly potent antibiotics, and anyone who has used these drugs will know of the inevitable stomach cramps and other gastrointestinal side-effects that are associated with their use.

The search continues for the ideal beta-lactam antibiotic, and numerous esoteric substances have been isolated from natural sources or have been synthesised during the past 15 years. These include the penems, carbapenems, cephamycins, monobactams, *etc*. The penems are wholly synthetic and from a clinical viewpoint have not been successful, while more success has been obtained with the carbapenems, which include the natural product thienamycin (from *Streptomyces cattleya*) and the analogues imipenem and meropenem. These carbapenems have a remarkable spectrum of antibacterial activity with a very high activity against pseudomonads. Thienamycin is relatively unstable, but the addition of a formamidine group to produce

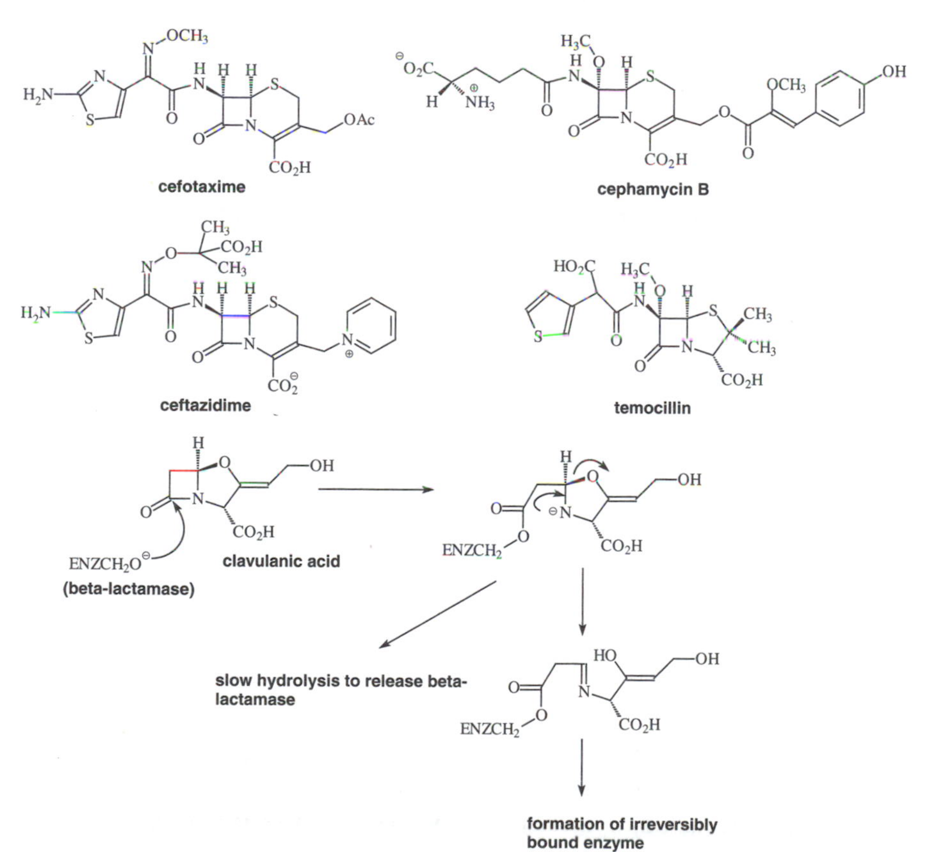

Figure 2.5

imipenem enhanced the stability greatly; however, the utility of this is compromised by its kidney toxicity. The introduction of a methyl group to produce meropenem eliminated the nephrotoxicity and maintained a high level of potency, and this drug is often given as a single injection prior to surgery. However, none of these compounds can be prepared by fermentation technology; thus, efficient syntheses are essential. The trinems are a further novel

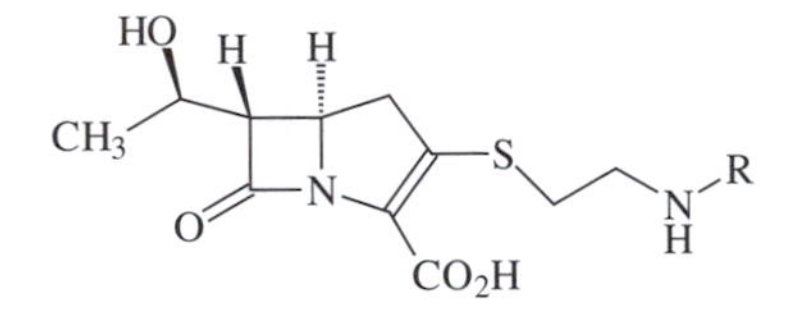

thienamycin (R = H)
imipenem (R = CH=NH)

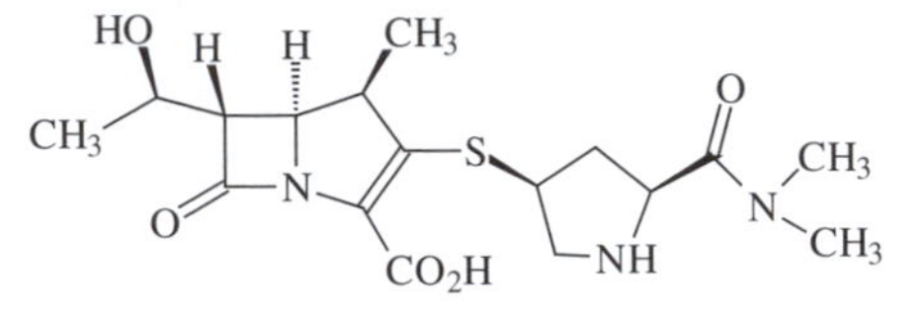

meropenem

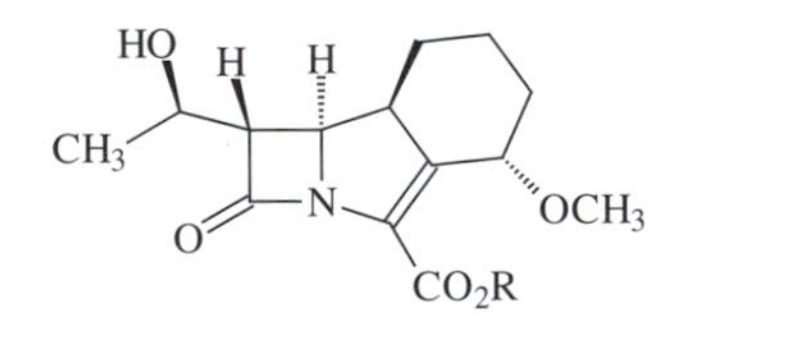

sanfetrinem (R = Na)
sanfetrinem cilexetil (R = CH_3 O O O

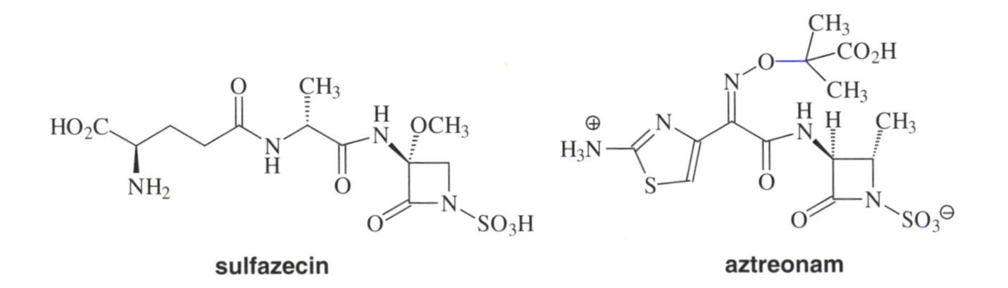

sulfazecin **aztreonam**

and totally synthetic class of beta-lactam antibiotics first reported in 1995 by the Glaxo-Wellcome laboratories in Verona. The archetypal compounds sanfetrinem and its prodrug form sanfetrinem cilexetil have an impressive spectrum of activity, including activity against penicillin-resistant *Streptococcus pneumoniae*. The cephamycins are typified by cephamycin C from

Streptomyces clavuligerus and this has a good spectrum of activity and good stability towards beta-lactamases. Various novel penicillins like temocillin also incorporate the alpha-methoxy group found in the cephamycins, and these are also resistant to the actions of the beta-lactamases. Finally, the monobactams typified by the simple sulfamate sulfazecin from *Chromobacterium violaceum* found in a New Jersey soil sample led the American company Squibb to produce the broad-spectrum antibiotic aztreonam (Azactam) in 1983. This has high stability against beta-lactamases and also especially good activity against pseudomonads.

However, none of these drugs has yet attained the star status of ampicillin, amoxycillin or cephalexin. It is also often forgotten that one of the best penicillins is the oldest one of all – penicillin G. It is unstable to acid, destroyed by beta-lactamases, but is nonetheless extremely effective (if given in large enough doses) against such rare conditions (at least in the developed world) as diphtheria, tetanus, anthrax, Lyme disease (a tick-borne disease) and venereal diseases. In addition, it usually produces few side-effects – and is thus a close approximation to a true 'magic bullet'.

THE WHITE PLAGUE AND ITS TREATMENT

Robert Louis Stevenson had it but survived; John Keats died from the disease (aged 26), so did D. H. Lawrence, the composers Chopin and Paganini, Balzac and most of the Bronte children. Tuberculosis or the 'white plague' has probably caused more suffering and premature deaths than any other disease in recorded history – an estimated one billion deaths in the past two centuries alone, accounting for about a quarter of all deaths in Europe and the USA during the 17th to 19th century. About one-third of the world's population harbour the causative bacillus – *Mycobacterium tuberculosis* – but do not develop the disease unless their immune system becomes weakened in some way. In countries where poverty and overcrowding are the norm, tuberculosis is common, and it causes as least three million deaths each year (that is, one death every 10 s!), making it the most important cause of death involving an infectious organism.

Like the common cold and influenza, it is a social disease, since the bacillus is transmitted by droplets carried through the air and in infected sputum. Carriers of the bacillus are generally unaware of its presence and show no symptoms of overt tuberculosis. They may, however, be identified by the tuberculin test, invented by Robert Koch in 1890. This involves impregnation of the skin with a small amount of heat-inactivated bacillus and observing the formation of a classic immune response.

Initial infection occurs when bacilli are taken up into the airways of the lungs. Here, they attract the attention of white blood cells, mainly phagocytes and macrophages, which engulf the bacilli but do not manage to kill them.

The bacterium multiplies and eventually destroys the macrophage, releasing more bacilli to be engulfed by other macrophages. After about two weeks, the immune system in the shape of thymocytes (T-cells) aggregate around the colonies of macrophages and 'wall them off', forming structures known as granulomas. The bacilli are thus killed or rendered dormant until the person's immune system is weakened by illness or old age. Patients with overt TB exhibit increasingly serious symptoms as the bacillus multiplies in the lymph nodes and then spreads to other parts of the body. The ultimate cause of death is usually gross destruction of the lungs, although the disease can affect almost any organ or part of the body. Other serious forms of TB are meningeal tuberculosis and Pott's disease, which involves damage to the spine. It is an ancient disease and there is considerable evidence of tubercular bone lesions in Egyptian mummies. The ancient Greek name for the condition was *phthisis*, although it has also been called scrofula, consumption and the 'King's evil'. This latter name arose because kings were supposed to inherit curative powers at the time of their annointment, and the merest royal touch on the swollen neck glands (lymph nodes) was believed to be efficacious.

Descriptions of the disease appear in the writings of almost every culture. John Bunyan, writing in the 17th century, claimed (in *The Life and Death of Mr. Badman*) that:

> The captain of all these men of death that came against him to take him away, was the Consumption for it was that that brought him down to the grave.

At that time, the disease probably caused about one-quarter of all deaths in England. The poets and writers of the so-called Romantic Age often seemed to be almost resigned to their fate. Keats mentions the symptoms in his poem *Ode to a nightingale*:

> The weariness, the fever, and the fret
> Here, where men sit and hear each other groan;
> Where palsy shakes a few, sad, last grey hairs,
> Youth grows pale, and spectre thin, and dies.

There were numerous (reasonably efficacious) remedies ranging from inhalation of the vapours of warm herbs and resins, through the widespread use of opium, to a late 19th-century preference for cod liver oil. The more wealthy sufferers were able to escape to the seaside or to the clean air of the mountains, where they stayed for months in the various sanitoria. But it was not possible to contemplate control of the disease until the causative organism had been identified. In 1865, the French army surgeon Jean-Antoine Villemin showed that phthisis was transferable from a tubercular animal to an unaffected animal by inoculation of infected tissue. However, the credit for the discovery of the bacillus is usually awarded to Robert Koch, who announced, in 1882, that he had been able to grow the organism in specially heat-treated ox or sheep sera. As already mentioned, he went on to invent the tuberculin test and even injected himself with a largish dose of heat-treated bacillus, which elicited a classic immune response: painful limbs, lassitude and a high temperature. All the glory went to Koch, causing Villemin to comment ruefully: "In science the credit goes to the man who convinces the world, not to the man to whom the idea first occurs."

The next major advance came with the production of a vaccine by Albert Calmette and Jean-Marie Guérin in 1924. They worked with a strain of tuberculosis isolated from cattle (*Mycobacterium bovis*), and this became less virulent on sub-culturing, but could nonetheless provide protection if injected into several different animal species. The vaccine was soon shown to be effective in humans, and was named after its inventors – Bacilli Calmette–Guerin or BCG – and has for many decades been given routinely and with great success as a prophylactic measure to children in most developed countries.

As for treatment of overt tuberculosis, this had to await the discoveries of another 'giant' of the antibacterial scene – Selman Waksman. He was born in July 1888 in the Ukraine, although his family migrated to the USA in 1910, and he received his scientific education in the Microbiology Department at Rutgers University in New Jersey. It was here that he acquired an interest in bacteriology, especially as it related to soil microorganisms, and after graduating in 1915, he took up a position as a research assistant at the New Jersey Agricultural Experimental Station. His work resulted in the isolation of a large number of moulds from the various soil samples, and he was sent to Charles Thom in Washington, to have them identified. This is another of those strange coincidences that abound in the history of antibiotic discovery, for this was the same scientist who was to be instrumental in identifying Fleming's mould, and in advising Florey and Heatley when they arrived in the USA in 1941.

In his autobiography, entitled *My Life with Microbes*, Waksman commented upon the importance of these formative years in bacteriology:

> I was thus impressed, at the beginning of my scientific career, by two important principles that were to serve as guiding lights in my whole future work, namely, the recognition that the soil is made up of a large number of different groups of microorganisms, each possessing different functions and activities, and that these microorganisms influence one another in a variety of ways.

In 1916, Waksman left New Jersey for the more pleasant climate of California, where he studied at Berkeley for his Ph.D. Here, he investigated the properties of soil microrganisms of the family Actinomycetes, whose filamentous structure resembles that of the fungi, but whose physiology and size are more like those of the bacteria. Upon completion of his Ph.D. course in 1919, he returned to the East coast to the Microbiology Department of the

New Jersey Agricultural Experimental Station of Rutgers University. With the exception of a ten-year period (1929–1939) during which he studied humus, he devoted the rest of his life to the Actinomycetes.

In 1939, his research group commenced an intensive programme that sought to identify soil microorganisms that were active against pathogenic bacteria. This type of systematic study had already been used to good effect by one of his former students, Rene Dubos, who had isolated the antibiotic gramicidin in 1939 from *Bacillus brevis*. This was a potent antibacterial substance but was too toxic for human use. Wakman's group quickly established that around 40% of their cultures had antibacterial activity and that about half of these were very potent. This confirmed his view that soil microorganisms probably played a large part in the 'purification' of soil that has been contaminated by, for example, human excreta. Their first major success was the antibiotic actinomycin A from *Actinomyces antibioticus*, and although this proved to be too toxic for human use, a related structure, actinomycin D (dactinomycin), did find limited use in cancer chemotherapy (see Chapter 4).

Waksman then turned his attention to the quest for an antibiotic to treat TB. The problem here was that *Mycobacterium tuberculosis* was very slow growing in culture – one cycle every 24 h in comparison with most bacteria, which have a reproductive cycle of 20–30 min – hence, testing the various soil samples for anti-mycobacterial activity would be a slow and tedious process. His son, Byron, at that time training to be a bacteriologist, suggested that he test his samples against the less pathogenic but much faster growing *Mycobacterium phlei*, and this provided a partial solution to the problem of biological evaluation. But how to identify the most promising strains of soil microbes? The solution to this problem was both neat and of potentially widespread utility: the soil samples were enriched with *Mycobacterium tuberculosis* and those actinomycetes that could produce anti-tubercular antibiotics obviously were the ones that thrived. With these two innovations in place, the scene was set for the discovery, in 1943, of the first major anti-tubercular drug – streptomycin from *Streptomyces griseus*, and the structure was determined in 1947 but not synthesised until 1974.

Clinical evaluation began in 1944 and streptomycin quickly attained the status of a miracle drug. Whereas patients with tubercular meningitis had previously had a zero chance of survival, as many as three-quarters of the patients could now be saved, and demand for the drug soon outstripped the rate of supply. By 1953, the rate of production in the USA had risen to 20,000 kg each month and there were eight companies in the USA producing the drug, with a further three in France, two in the UK, four in Japan and two in Italy.

There were a few problems when streptomycin was given in large doses, and the most important of these was damage to the aural nerve with resultant deafness. But it was the first effective drug against TB and is still used for this purpose, although it has largely been superceded by other safer drugs. It is also the drug of choice for the treatment of *Yersinia pestis* (plague), and in combination with penicillins or vancomycin, for the treatment of streptococcal endocarditis.

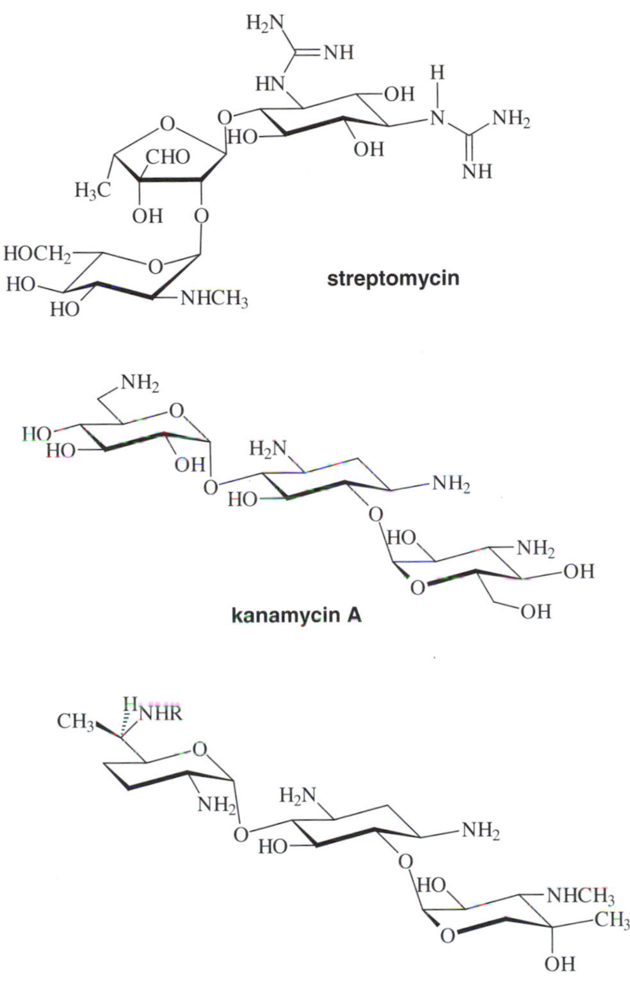

streptomycin

kanamycin A

gentamicin C_1 (R = CH_3) and C_2 (R = H)

This was not the end of the story, and over the years, several other antibiotics of the same structural type as streptomycin were discovered and subsequently entered clinical practice. These include: neomycin (1949) from *Streptomyces fradiae* (especially for dermatological infections); kanamycin (1957) from *Streptomyces kanamyceticus* (initially used for infections caused by pseudomonads – organisms that are especially dangerous for burn victims – but has now been replaced by safer drugs); gentamicin (1963) from

Micromonospora purpurea (especially for Gram-negative bacilli); and tobramycin (1967) from *Streptomyces tenebrarius* (also for Gram-negative bacilli). All these are now called *aminoglycoside antibiotics* and their mode of action involves inhibition of protein biosynthesis after binding to the 30S subunit of the bacterial ribosomes. These are the complex cytoplasmic structures on which protein biosynthesis takes place, and the drugs interfere with the translation process whereby m-RNA directs the incorporation of amino acids into the growing polypeptide chain, thus inhibiting the production of the bacterial proteins (see Fig. 4.3). Since they have no great affinity for mammalian ribosomes, they have a good degree of selectivity and thus satisfy at least some of the criteria required of 'magic bullets'.

All these aminoglycoside antibiotics (and many other antibiotics discovered in more recent times) owe their discovery, in no small part, to Selman Waksman's enthusiasm for soil microorganisms. He richly deserved the Nobel Prize for Physiology or Medicine awarded to him in 1952; and at that time, it seemed perfectly reasonable for him to predict (in his autobiography):

> It (streptomycin) has opened a new field of therapy, it has pointed a way towards the final solution of tuberculosis, whereby this disease as well as syphilis and pneumonia will pass into the limbo of history.

His optimism has, unfortunately, not been justified.

In the same year that Waksman received his Nobel Prize, another major anti-tubercular drug was introduced. This was the totally synthetic isoniazid, which arose out of studies carried out independently by groups at Hoffmann LaRoche in Nutley, New Jersey, and at the Squibb Institute in New Brunswick, Canada. Both groups were attempting to modify the biological properties of a group of compounds known as thiosemicarbazones. Some years previously, Domagk, discoverer of the sulfonamides, had shown that one such compound, thiacetazone, had some anti-tubercular activity. The Hoffmann LaRoche and Squibb scientists were trying to prepare isonicotinaldehyde thiosemicarbazone, which had been chosen as a target because it possessed the basic skeleton of the B vitamin nicotinamide. This vitamin had also been shown (by a group at Lederle in 1948) to have anti-tubercular activity. In the event, the key chemical intermediate, isoniazid, was shown to have very potent anti-tubercular activity in its own right. In fact, it was 15 times more potent than streptomycin and much cheaper to produce. Not surprisingly, it soon became the most important drug for the treatment of TB, and remains an important component of the cocktail of drugs that is still given to patients.

The other major drug in first-line therapy is rifampicin (Rifampin), which is a chemically modified form of the antibiotic rifamycin B. This was first isolated in 1957 by a group at Lepetit in Milan from a strain of *Nocardia mediterranei* (also part of the family Actinomycetes) from a soil sample taken from the Côte d'Azur. Its structure was elucidated by Prelog and Oppoltzer in 1963 and later confirmed by X-ray crystallography. The name has an interesting aetiology since it is derived from the famous Jules Dassin film *Rififi*, which was popular at the time. The term *rififi* comes from French slang common in the backstreets of Marseilles and means 'fisticuffs'. The native antibiotic, rifamycin B, was neither particularly potent nor very stable. In fact, oxidation by the air provided a more potent and stable species, rifamycin S, and this in turn was converted chemically into the semi-synthetic compound rifampicin. This possessed marked anti-tubercular activity, and was also highly effective against the bacillus, *Mycobacterium leprae*, that causes leprosy. It was introduced as a drug in 1966 and has proved its worth against both TB and leprosy during the past four decades.

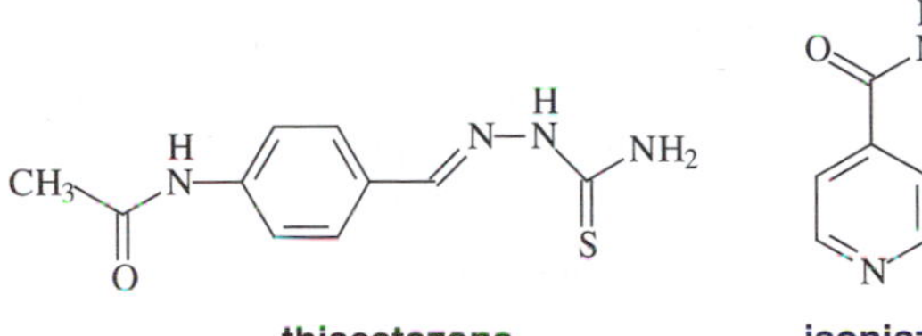

thiacetazone

isoniazid

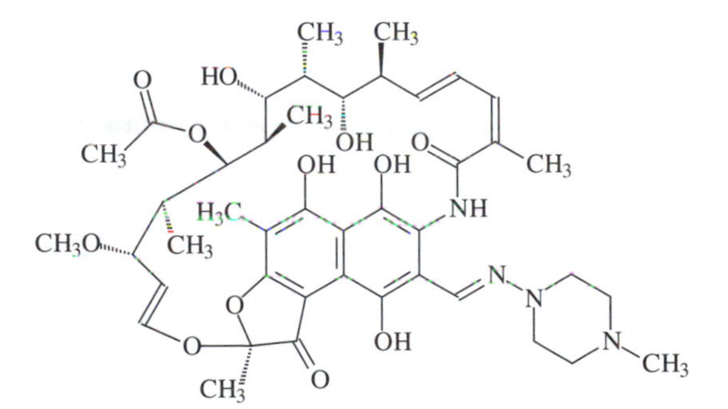

rifampicin

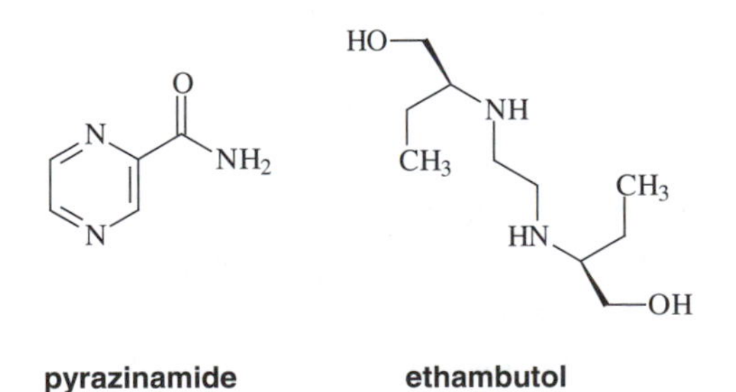

pyrazinamide **ethambutol**

With this veritable armoury of drugs and prophylactic vaccination with BCG, tuberculosis seemed doomed to extinction as Waksman had prophesied.

Indeed, the level of overt disease in the developed countries plummeted between 1950 and 1960 and then fell more slowly until the mid-1980s, eventually levelling out at around 10 cases per 100,000 of the population. Since then, the incidence has risen slowly but inexorably at a rate of about 5% per annum and as much as 13% per annum in places like New York City. The reason for this is not hard to discern. Over the years, especially in developing countries, thousands of patients have been treated with streptomycin, rifampin, isoniazid and other drugs, but have failed to complete their course of treatment. This is hardly surprising since this can take a minimum of six months and is obviously very expensive. These patients achieved a partial cure for their disease but still harboured *Mycobacterium tuberculosis*, and additionally the residual bacilli had very likely mutated in response to prolonged exposure to the drugs to produce drug-resistant strains. Couple this with the rapid growth in cheap air travel during the past three decades and the recipe for disaster is all too apparent. But the situation has been exacerbated since the mid-1980s by the growing incidence of HIV and associated AIDS (to be discussed in detail in Chapter 3). AIDS patients have seriously weakened immune systems, and since many of them also harboured *Mycobacterium tuberculosis*, it was almost inevitable that they would develop overt tuberculosis. Whereas HIV has a relatively inefficient (sexual) route of transmission, the bacilli are easily passed on through everyday social contacts. In 1983, the World Health Organisation (WHO) declared a 'global emergency' for TB. It estimates that worldwide there are presently eight million new cases of TB each year and up to three million deaths. Figures released by the WHO in 2003 suggest that 60% of the TB patients in sub-Saharan Africa are co-infected with HIV, and in Asia the figure is 35%. What can be done?

While the situation is now very serious, there is a glimmer of hope. In recent years a great deal has been learnt about the chemical structure of the cell wall of the bacillus (Fig. 2.6). This is much more complex than that of the Gram-positive bacteria, which are relatively easy to treat with conventional antibiotics. In addition to an inner layer of peptidoglycan (similar to that of the Gram-positive organisms), it has a highly complex outer layer comprising a combination of unique lipids (mycolic acids) and a polysaccharide composed of galactose and arabinose (arabinogalactan) both in the unusual furanose form. The arabinogalactan is cross-linked to the peptidoglycan via a unique disaccharide comprising rhamnose and *N*-acetylglucosamine. This structure denies access to many conventional antibiotics, but its unusual chemical constitution may offer an opportunity for the drug designers. Since the mycolic acids are unique to the mycobacteria, drugs that interfere with their production should not affect mammalian biochemistry. These mycolic acids are alpha-branched beta-hydroxy fatty acids with 60–90

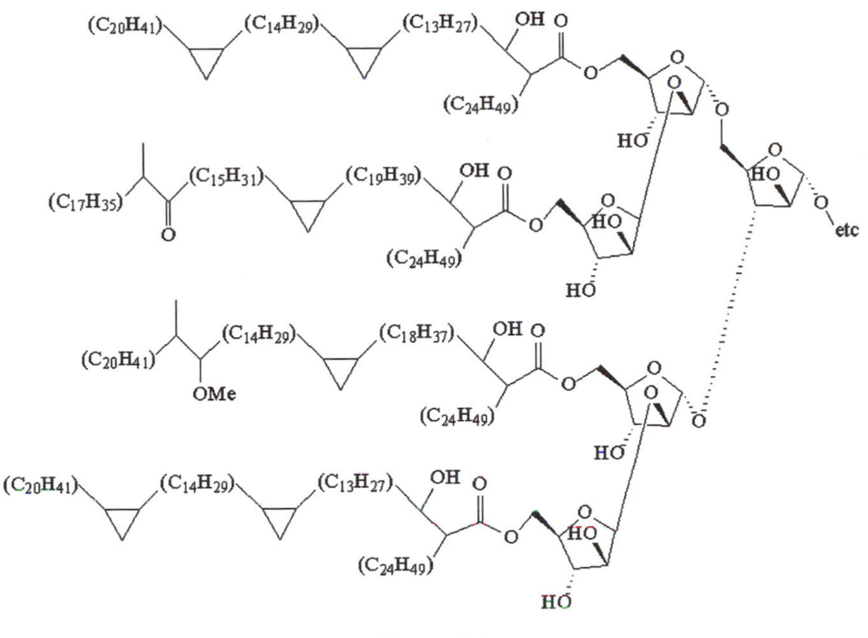

Figures 2.6

carbon atoms and many cyclopropyl rings. In addition, the sugar arabinose is not used in mammalian metabolism; thus, drugs that inhibit the production of arabinose or its combination with the mycolic acids may also show efficacy. Indeed, the drug ethambutol does inhibit the incorporation of arabinose units into the arabinogalactan, and isoniazide inhibits the biosynthesis of mycolic acids. As for the mode of action of streptomycin, it is believed to displace polyamines like spermine and spermidine, which are also key constituents of the mycobacterial cell membrane. Like the other aminoglycosides, streptomycin also disrupts protein biosynthesis through binding to the 30S ribosomal subunit.

In the meantime, the TB patient must endure six months of chemotherapy. The regime usually involves a drug cocktail including rifampicin, isoniazid and pyrazinamide (another synthetic anti-tubercular drug), together with either ethambutol (also synthetic) or streptomycin for two months. If the patient appears to be responding, the drugs isoniazide and rifampicin are given for a further four months or until a complete cure has been achieved. Whenever possible, a regime known as DOTS – directly observed therapy short-course – is imposed to ensure that the patient complies with the full course of treatment. For refractory TB, other drugs are used, including the aminoglycosides kanamycin and amikamycin (introduced in 1972), and quinolone antibacterial agents like ciprofloxacin (which will be discussed later in this chapter).

In the longer term, it may be possible to design drugs that will stimulate the weakened immune system to eradicate the bacilli. Some work has also been done to modify the potency of BCG, and to produce 'naked DNA' vaccines. This strategy involves vaccination with small pieces of DNA that code for particular proteins of the bacillus cell wall. Interestingly, the genome sequence of the bacillus is rich in genes coding for the production of enzymes of lipid metabolism (mainly those for beta-oxidation), and this implies that the bacilli literally dine on the patient's lipids. But for the moment, Waksman's optimism for the complete eradication of tuberculosis by chemotherapy must be considered premature.

MORE 'TREASURES' FROM STREPTOMYCES

Selman Waksman was not alone in his enthusiasm for screening soil samples for antibacterial activity. In 1944, the Lederle company of Pearl River, Mississippi, USA, employed the retired mycologist Benjamin Duggar to advise them on a screening programme that was seeking new anti-tubercular agents. He solicited samples from friends and ex-academic colleagues from all over the world, and in 1945 one of these, from the state of Missouri, proved to be particularly interesting. The golden-yellow culture, which he christened *Streptomyces aureofaciens*, provided a novel antibiotic that became known as chlortetracyclin (Aureomycin). This had broad-spectrum antibacterial activity, and went into production using the type of large-scale fermentation technology that had been developed for the penicillins. It reached the market in 1948.

One year later, the American pharmaceutical company Pfizer discovered a related structure – christened oxytetracycline (Terramycin) – from *Streptomyces rimosus*. Interestingly, this was found in a soil sample located near their factory in Terre Haute, Indiana. The parent structure – tetracycline – was then obtained by chemical removal of the chlorine atom (an element only rarely found in terrestrial organisms but common in natural products from marine organisms) from chlortetracycline. This third antibacterial agent was subequently found naturally as a constituent of both *Streptomyces aureofaciens* and *Streptomyces viridifaciens*. The structures of chlortetracycline were established by R.B. Woodward in 1952 and that of oxytetracycline by Pfizer scientists (in collaboration with RBW) in 1952.

All these antibacterial agents were orally active (unlike penicillin G), had broad-spectrum activity and were cheap to produce. They quickly became very popular for the treatment of a wide variety of bacterial infections, although they can produce unpleasant side-effects – like colonic bleeding – upon prolonged use. This is due to their poor absorption through the gut wall and consequent presence in the intestines, where they damage the natural

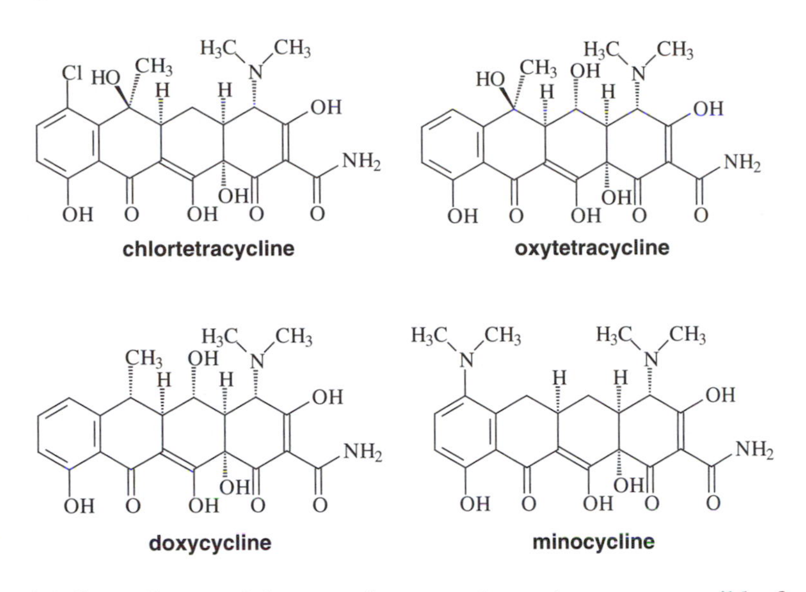

bacterial flora. Some of these endogenous bacteria are responsible for the synthesis of species like vitamin K, which is a key component of the blood clotting process – hence the loss of blood.

Notwithstanding these side-effects, the drugs, and several more recently prepared semi-synthetic analogues, are still the agents of choice for the treatment of cholera, Lyme disease (a tick-borne disease), plague, typhus, Rocky Mountain spotted fever, non-specific urethritis and chlamydial infections like trachoma and psittacosis. In addition, the drug doxycycline has joined the armoury of prophylactic medicines for malaria, and minocycline has an important role in the treatment of meningococcal meningitis, since its enhanced lipophilicity allows its uptake into the cerebrospinal fluid. Their mode of action is related to that of the aminoglycosides, in that they bind to the 30S ribosomal subunit and prevent binding of the aminoacyl transfer-RNAs, thus inhibiting bacterial protein biosynthesis. They have the same mechanism of action in mammalian cells, but since they are not so readily taken up by these cells, they have a fair degree of selectivity for bacterial cells. Like the sulfonamides they are primarily bacteriostatic rather than bactericidal, and reversibly inhibit the growth and multiplication of susceptible bacterial species rather than killing them outright.

Almost contemporaneous with these discoveries, scientists at Parke Davis in Detroit were also studying soil samples from all over the world. These were evaluated at Yale University, and in 1947 they isolated a novel structure from a soil sample taken from near Cararcas in Venezuela. This was, unsurprisingly, christened *Streptomyces venezuelae*, and the antibiotic was given the name chloramphenicol. Its relatively simple chemical structure was elucidated in 1947, and it was also first used in that year to treat 22 seriously ill patients in a typhus outbreak in La Paz, Bolivia. All survived, and the initial success of the drug was ensured. Unlike most of the other drugs mentioned so far, chloramphenicol was easy and cheap to prepare in a few chemical steps, and rapidly became Parke-Davis's best-selling drug, worth almost $10 million in 1949. Then disaster struck with the revelation that a tiny proportion of people (no more than 1 in 20,000 and perhaps as few as 1 in 100,000) contracted aplastic anaemia when treated with the drug. Sales dropped precipitously as GPs moved their allegiance to other companies and their antibacterial products. These days, the drug is usually reserved for external use in the treatment of ear and (in particular) eye infections. The mode of action is very similar to that of the tetracyclines, but this time the antibiotic binds to the other main ribosomal subunit (50S subunit), where it inhibits the enzyme peptidyl transferase, and disruption of bacterial protein biosynthesis results.

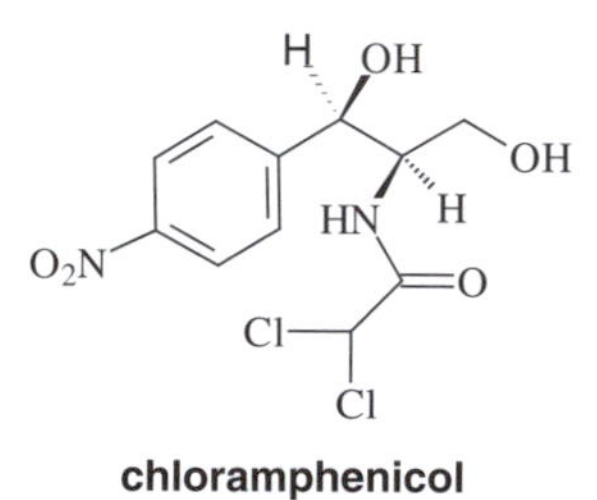

chloramphenicol

The US pharmaceutical company Eli Lilly also had its own screening programme, and in 1952 isolated the first member of a new family of antibiotics from *Streptomyces erythreus* found in a soil sample sent from the Philippines. This novel antibiotic, given the name erythromycin, had a spectrum of antibacterial activity similar to antibiotics like penicillin G and V, but was extremely acid-sensitive and relatively insoluble in water. When its chemical structure was elucidated in 1957, the antibiotic was revealed to comprise a large ring lactone with two unusual carbohydrate units attached to it. This new structural type was given the name *macrolide* antibiotic and, like chloramphenicol, was shown to exert antibacterial activity by binding to the ribosomal 50S subunit and inhibiting bacterial protein biosynthesis.

Erythromycin has a relatively broad spectrum of activity and its major uses these days are in the treatment of various types of bacterial pneumonia (including the notorious Legionnaires' disease), and the severe gastrointestinal effects due to *Campylobacter jejuni*. Its instability in the presence of gastric acid has been overcome by administering the drug as a coated tablet, which aids its survival in the stomach and increases water solubility. It is a relatively safe drug, although it does cause some gastrointestinal discomfort. More recently introduced members of the family include azithromycin and clarithromycin and these are less acid-sensitive and also have greater bioavailability.

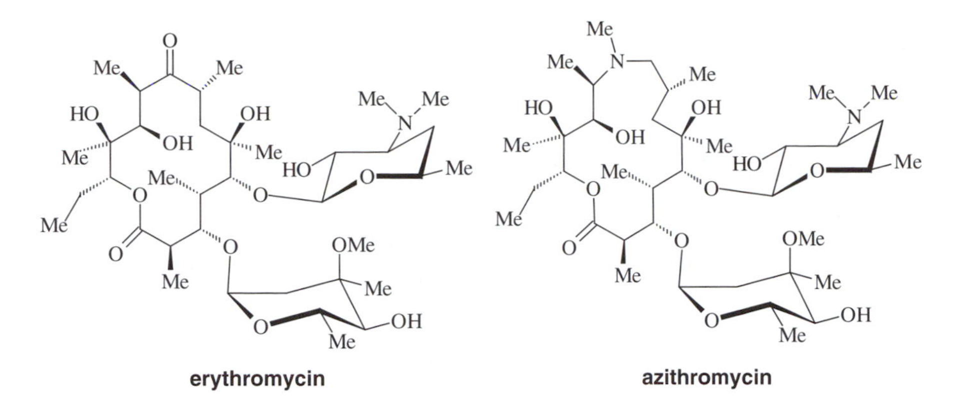

In recent years, it has become important for another reason. Several research groups, comprising chemists and molecular biologists, have managed to unravel the intimate details of the pathway of biosynthesis of erythromycin. This has involved identification of the bacterial genes and the enzymes for which they are the blueprint. But the particularly exciting part of this work has been the use of this knowledge to produce mutant strains of *Saccharopolyspora erythrea* (another producer of erythromycin), and other organisms in which the genes have been rearranged. Not surprisingly, many of these mutant strains are non-viable, but some of them produce modified chemical structures (rather than erythromycin), where the structural changes are exactly what one would predict from the gene manipulation. Similar studies with other microorganisms and the natural products that they produce have not only provided a wealth of knowledge about the genetics of biosynthesis of these compounds but also a range of new 'unnatural products'. The potential for the production of 'designer antibiotics' through genetic manipulations of this kind are quite awesome, although it must be admitted that these studies are only at an early stage at present (see Figs. 2.7 and 2.8).

Lilly's contribution to the soil-derived antibiotic arsenal also extended to the discovery of vancomycin, from *Streptomyces orientalis* of Indian and Indonesian origins. This complex antibiotic was first introduced in 1956 and was valued for its marked effect on penicillin-resistant staphylococci. It has largely retained this superiority over these organisms and is now often the agent of last resort in patients who have contracted MRSA infections.

It shares its mode of action with the antibiotic eremomycin, discovered in the late 1980s. Both antibiotics associate to form dimers that become attached to the bacterial wall by binding to the D-alanine–D-alanine moieties of the peptidoglycan. These associations have been demonstrated by extensive structural studies carried out by Dudley Williams and his group at Cambridge, and involve a precise alignment of the antibiotic with the D-alanine–D-alanine unit of the bacterial peptidoglycan precursor (Fig. 2.9). The drugs thus deny

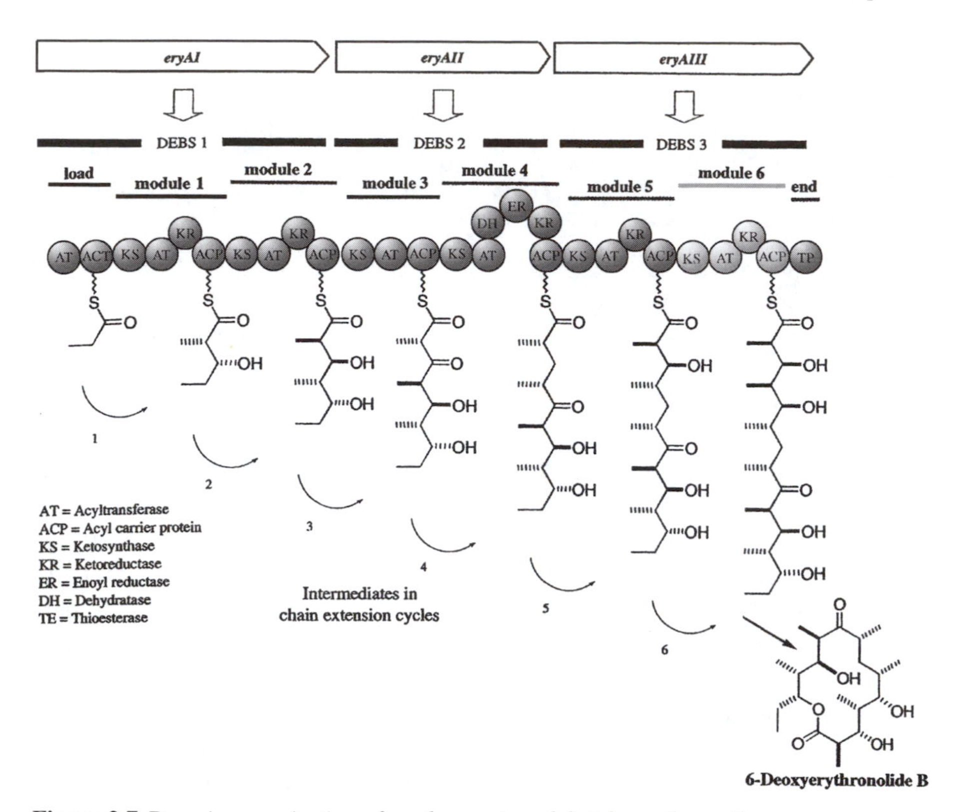

Figure 2.7 *Domain organisation of erythromycin polyketide synthase. Gene sequence putative domains are represented as circles and the structural residues are ignored. Each module incorporates the essential KS, AT and ACP domains, while all but one include optional reductive activities (KR, DH, ER). The one-to-one correspondence between domains and biosynthetic transformations explains how programming is achieved in this modular PKS. DEBS=deoxyerythronolide B synthase (reproduced with permission of Prof. James Staunton)*

access to the transglycosylase (which catalyses the synthesis of the polysaccharide) and transpeptidase enzymes (which catalyse the cross-linking step), and thus inhibit the formation of a cross-linked cell wall peptidoglycan, and the bacteria die. The related antibiotic teicoplanin, from *Actinoplanes teichomycetius*, does not form dimers, but does have a long hydrocarbon tail and this helps it to associate with the bacterial cell membrane prior to the association with the D-alanine–D-alanine moiety. Very recently, Eli Lilly has produced an analogue of eremomycin, which also has a lipophilic side-chain, and this semi-synthetic drug is 1000 times more potent than vancomycin, presumably because it too can bind very readily to the cell membrane.

For some reason, which is presently not understood, it has not been possible for most bacteria to develop means of overcoming the effects of these antibiotics; thus, resistance to vancomycin and teicoplanin has been, until recently, quite rare. This appears to be due to the fact that as many as nine genes are required to overcome the effects of vancomycin. Two of the required genes

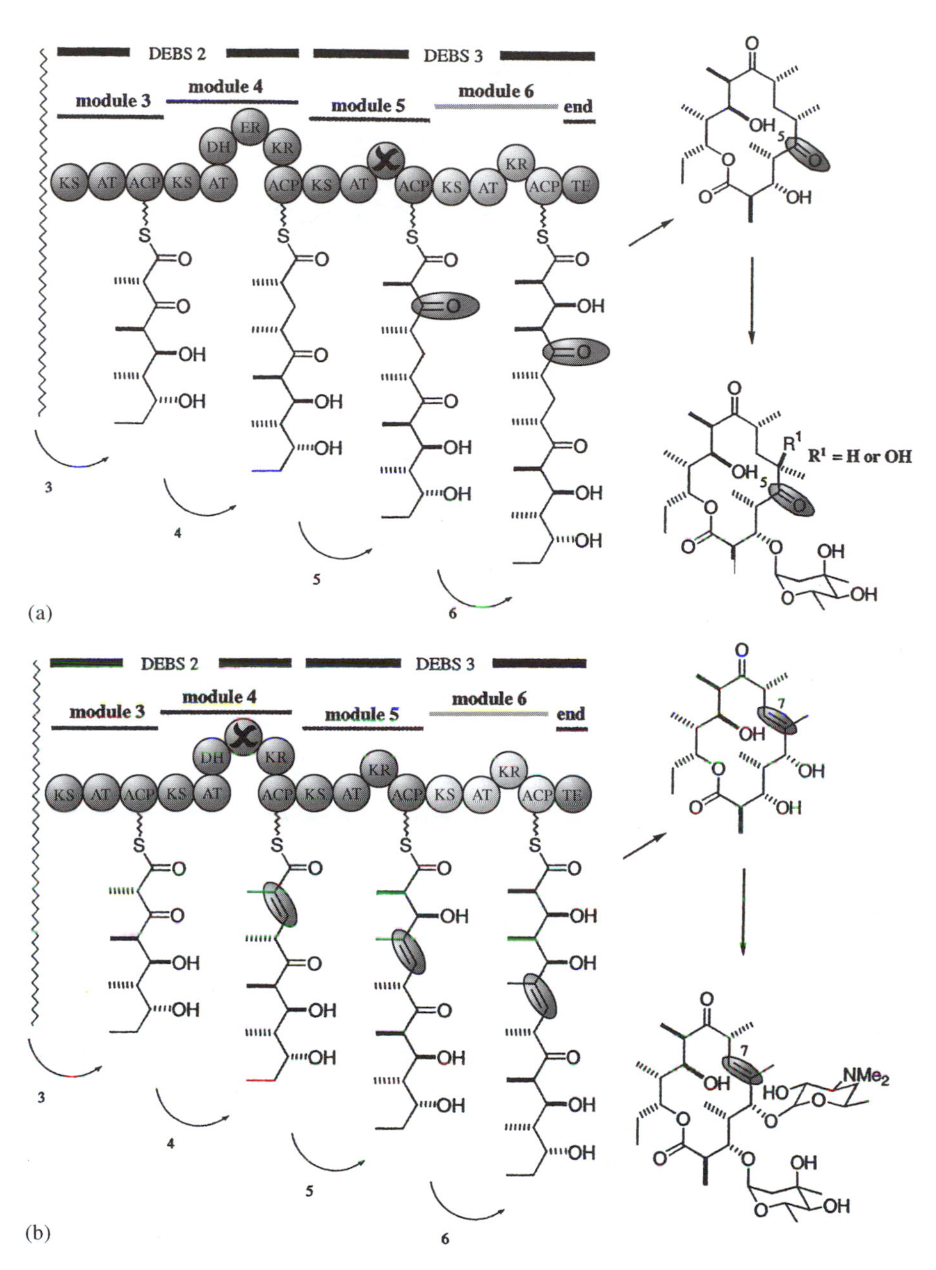

Figure 2.8 *Experimental evidence for the modular organisation of DEBS. (a) Inactivation of KR_5 of DEBS resulted in the production of erythromycin analogues with keto groups at the C-5 position. (b)Inactivation of ER_4 resulted in an analogue of erythromycin with a double bouble bond at the expected site (reproduced with permission of James Staunton).*

apparently code for proteins that enhance the transfer of the resistant genes between bacteria. Another codes for a reductase enzyme that reduces pyruvic acid to lactic acid and a further gene codes for a protein that couples D-alanine

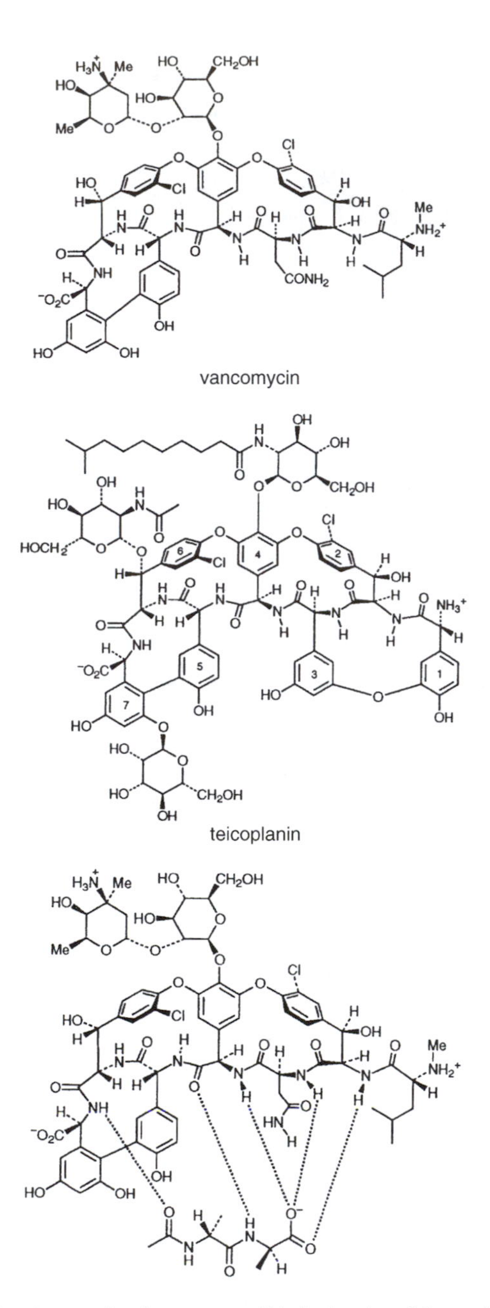

Figure 2.9 *The binding interaction between antibiotic (top) and bacterial cell wall analogue N-acetyl-D-Ala-D-Ala (below). Hydrogen bonds between the two are represented by dotted lines. (Reproduced with permission of Professor Dudley Williams).*

to D-lactic acid – and these are critical since resistant bacteria have a cell wall peptidoglycan with D-alanine–D-lactic acid moieties instead of the usual D-alanine–D-alanine termini. Despite these obstacles to bacterial resistance, a growing number of hospitals around the world have begun to report the existence of vancomycin-resistant enterococci (VRE), especially in surgical units. The days of these 'last-resort' antibiotics may be numbered.

THE RISE OF THE 'SUPERBUGS'

For most people, the death of Jim Hensen, creator of the Muppets, from a rampant staphylococcal infection in the spring of 1990, was the first time that they became aware of 'superbugs'. For 45 years, antibiotics had been the miracle drugs that kept them free of the disease and premature death that had afflicted their forbears. Now, that had all changed. Throughout the 1980s, an increasing number of cases of what became known as 'toxic shock syndrome' were reported. This was apparently caused by a new strain of *Staphylococcus aureus* that released toxins into the bloodstream. On first contact, the infected person became sensitised to the toxins, and if they were re-infected, a fulminating immune reaction was initiated, and this filled the lungs, kidneys and other organs with antibody–toxin products. The patient was often overwhelmed within a few hours, and there was no time for antibiotic therapy to be effective.

The most infamous examples of this syndrome followed the introduction of a new, highly absorbent tampon in the early 1980s. These became popular because they could be left in place for many hours, and women unknowingly incubated the new virulent strain of staphylococcus under almost optimum growth conditions. During the first infection, sensitisation to the bacterial toxins occurred, and during a second or third menstrual period, a fulminating immune response was elicited. By 1984, over two and a half thousand American women had been treated for tampon-associated toxic shock syndrome, and 5% of these had died. The condition was also especially prevalent in Scandinavia and Germany, but occurred in almost every developed country.

Other strains of *Staphylococcus aureus* produce toxins that cause damage to the gastrointestinal tract (enterotoxins) or to the skin (dermotoxin or exfoliative toxin). These latter toxins are associated with the sensational headlines seen in recent years that report attacks by 'flesh-eating superbugs'. Less often reported by the media is the damage caused by other strains of staphylococcus like *Staphylococcus epidermidis* and *S. saprophyticus*, which are termed coagulase-negative organisms, so-called because, unlike most staphylococci, they do not cause blood clotting. They are, however, responsible for a majority of infections in the vicinity of medical implants like heart valves, hip joints and catheters used during renal dialysis. These infections range in severity from mild to life-threatening.

The existence of rogue strains of bacteria that were resistant to antibiotics was nothing new. As mentioned earlier in the chapter, by the early 1960s, 10% of the strains of *Staphylococcus aureus* produced penicillin-destroying enzymes (penicillinases) and were thus resistant to the first-generation penicillins. Today, the figure approaches 100%. The advent of methicillin and other penicillinase-resistant semi-synthetic penicillins provided a new lease of life for this class of drugs but the bacteria soon responded by producing altered penicillin-binding proteins in their cell walls. Today, one of the most dangerous features of a stay in hospital is the risk of infection with a strain of methicillin-resistant *Staphylococcus aureus* or MRSA.

How do the bacteria manage to overcome these one-time 'magic bullets'? We can easily appreciate why they have the ability to mutate, to counter these man-made toxins, when we recall the origins of most of our antibiotics – the soil microorganisms and airborne moulds. Bacteria have been co-evolving with these largely non-pathogenic organisms for at least three billion years, and have been forced to devise means of detoxifying the chemicals that they encountered. The mechanisms were thus already in place, even before man exploited what was the unfair advantage of chemistry, to modify the natural toxins. Faced with these novel, modified toxins, the bacteria have used three main strategies for overcoming them. They have mutated to produce new strains that have a modified cell wall structure so that the antibiotics can no longer penetrate into the cytoplasm of the bacterial cell. Some have also devised means of ejecting antibiotics (using efflux pumps) that do manage to penetrate the cell wall. In addition, mutant strains have appeared, which possess altered ribosome structures so that the tetracyclines, for example, can no longer attach to these structures and prevent protein production. Alternatively, some strains have key enzymes that have altered affinities for the antibiotics. For example, there are resistant strains of *Escherichia coli*, the common gastrointestinal bacterium, that have a new form of the enzyme dihydrofolate reductase that has an affinity for trimethoprim around 20,000 times lower than the normal enzyme. The drug can thus no longer inhibit this enzyme and the bacteria grow unchecked. Of considerable concern at the present time is the emergence of bacterial strains that have changed the composition of the building blocks of their coat. These no longer have the D-alanine–D-alanine dipeptide moiety but have mutated to use a D-alanine–D-lactate moiety; thus, vancomycin and teicoplanin – the antibiotics of last resort – cannot attach to this sequence (they typically have a 1000-fold drop in affinity) and thus have no effect on these vancomycin-resistant (gastrointestinal) enterococci (VRE).

However, by far the most important strategy involves the production of enzymes that destroy or deactivate the antibiotics. Bacterial enzymes that inactivate chloramphenicol (*e.g.*, chloramphenicol acetyl transferase) and the aminoglycosides (usually enzymes that acetylate or phosphorylate the aminoglycosides) are quite common, but it is the penicillinases or, as they are now called, the *beta-lactamases*, that have had the most impact on antibacterial chemotherapy. The beta-lactam antibiotics, including the penicillins, cephalosporins and the newer variants called penems, carbapenems, cephamycins and monobactams, all possess the four-membered ring (beta-lactam ring) that is the key to inactivation of the main bacterial cell wall building enzyme – the transpeptidase (see Fig. 2.4). These beta-lactamases destroy the beta-lactam ring and thus deactivate the antibiotics, and the mechanism shown in Fig. 2.5 is typical. The Glaxo Smith-Kline drug Augmentin (originally developed by Beecham) has become highly successful because its clavulanic acid component destroys the beta-lactamase, thus allowing its amoxycillin component to kill the bacteria in the usual fashion. A large number of beta-lactamase variants are known, and even a single base change in the gene coding for beta-lactamase can have far-reaching effects on substrate specificity, and these are most common in enteric bacteria. It is not uncommon for a resistant bacterial strain to combine a low-affinity penicillin-binding protein with a cell membrane that only allows reduced uptake of the beta-lactam antibiotics, with the addition of a highly potent beta-lactamase enzyme. It is a sobering thought that such simple changes in specificity can overwhelm an antibiotic that may have cost $100 million and ten years to develop.

Despite a huge investment of money and effort, the bacteria have managed to counter the early advantages of most of the beta-lactam antibiotics. The fact that *Staphylococcus pneumoniae* still kills 40,000 persons each year in the USA, or that most hospitals around the world report the existence of MRSA strains, leaves little room for complacency. So how have the bacteria achieved this remarkable near-supremacy over man's new miracle drugs?

The pathogenic bacteria have changed their genetic constitution millions of times during the aeons in which they have coexisted with other organisms. Most of these mutations give rise to a non-viable strain, but every so often, one of them will be not only viable but also more suited to its environment. During these millions of years of evolution, a pool of genes that confer resistance to the effects of the natural antibiotics have been produced and have been retained. When the pathogenic bacteria first encountered the man-made penicillins, tetracyclines, aminoglycosides, *etc*., they had to evolve new mechanisms and enzymes to cope with them or become extinct. If the new antibiotics had been used in a conservative manner, it is likely that the pathogenic bacteria would have evolved slowly in response to this new assault. Unfortunately, the antibiotics have often been prescribed in a most cavalier way. For example, they were often prescribed for the common cold (to keep the patient quiet!), and since this is caused by viruses, antibiotics have no immediate benefit. They were also incorporated into animal feedstuffs in order to improve the yields of, in particular, 'battery animals'. The pathogenic bacteria thus encountered these new drugs on a worldwide basis and at a frequency that required a rapid response. This strong selective pressure ensured that mutations occurred in existing genes for antibiotic resistance, hence the almost immediate appearance of strains with potent penicillinase activity. But more insidiously, the bacteria developed their capabilities in the transfer of genes between strains and indeed different organisms to pass on genes that coded for beta-lactamases, aminoglycoside-modifying enzymes, *etc*. These genes are transferred from one bacterium to another by means of *resistance plasmids*, which are small packages of DNA. It has been suggested that the R-plasmids have evolved during the past six decades – that is, during the antibiotic era, although the bacteria have probably always possessed discrete segments of DNA capable of moving from one gene to another – so-called *transposons*. As the bacterial genes mutated in response to environmental pressures including their exposure to the new antibiotics, these transposons appear to have aggregated via transposition to form *replicons* that became capable of autonomous replication. Ultimately, these replicons became the R-plasmids that are transferred to other bacterial cells.

The actual transfer takes place, most commonly, by means of what is known as *conjugation*, whereby small, hair-like projections on the surface of the bacterium make contact with another bacterium and the R-plasmids pass through these *sex pili* (*pilum* being the Latin for spear) from one organism to its neighbour. Introduction of the plasmids can also occur by a process called *transduction*, whereby the genetic material passes into the cell while it is incorporated with the genes of a bacterial virus (a bacteriophage). These means of gene transfer explain why the use of antibiotics in animal husbandry was so disastrous. The swapping of resistance genes occurred in the farmyard and the same genes passed into the gastrointestinal tracts of the human consumers when they ate inadequately cooked meat. It is probable that the serious outbreak of *E. coli* food poisoning, which occurred in Scotland in 1997, was caused by a strain of the bacterium that had acquired a toxin-producing gene from *Shigella* species that are normally responsible for bacterial dysentry. It is almost as if bacteria 'talk to one another' when they share a common environment, a person's gastrointestinal tract, for example. This type of 'conversation' involves a swapping of genes, however, rather than gossip!

Once the genes for antibiotic resistance were in the human line, their transmission around the world was ensured by the availability of cheap air travel. A sample of antibiotic-resistant *E.coli* food poisoning can be transported from Tokyo to London in a mere 12 hours! This situation has been grossly exacerbated by the ready availability of antibiotics 'across the counter' (rather than on prescription) in most developing countries, but also in places like the Middle East, and somewhat surprisingly, the USA. Additional problems have been caused by the common circumstance of people failing to complete the prescribed course of treatment because they feel better, and they stop taking the antibiotics before all the bacteria have been killed. Not surprisingly, the residual bacteria are the most rugged strains and these then transfer their resistance genes to the natural gut bacteria, where they form a reservoir of resistant genes for onward transfer to other (pathogenic) bacteria. A possible doomsday scenario where everyone harbours bacteria with a veritable cocktail of resistance genes is thus almost inescapable.

The pharmaceutical industry has responded to this threat by expanding their screening programmes for natural antibacterial substances and also by preparing a range of totally synthetic drugs. Of the latter type, the fluoroquinolones have been particularly successful. These emerged out of studies carried out at Sterling-Winthrop in the early 1960s on two totally synthetic compounds called nalidixic acid and cinoxacin. These possessed modest antibacterial activity, and a major programme was initiated to synthesise more potent analogues of these lead compounds. The first clinically useful fluoroquinolone, norfloxacin, was produced in Tokyo by the Kyorin Pharmaceutical Company in 1978; but it was the Bayer drug ciprofloxacin, first patented in 1982, that has proved to be the most successful of this new class of antibacterial drugs. In 1995, worldwide sales were worth $1.25 billion, which made it the second best-selling drug of all. It has broad-spectrum antibacterial activity against both Gram-positive and Gram-negative organisms, and is particularly valuable because it, and the other fluoroquinolones, have a novel mode of biological activity. The bacterial DNA is usually found

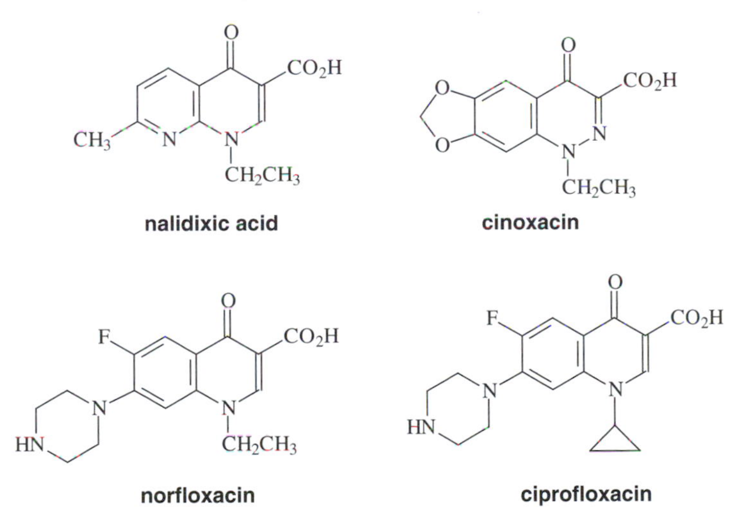

in a supercoiled state and requires enzymes called DNA gyrases to partially unravel this structure before replication and transcription can take place. The fluoroquinolones prevent these gyrases from functioning both at the unravelling stage and at the subsequent supercoiling stage. Since mammalian DNA exists in a different supercoiled state, and is processed by discrete enzymes called topoisomerases (to be discussed in Chapter 4), the fluoroquinolones have selective activity against bacteria. However, several strains of bacteria have been discovered that are resistant to these drugs, either because they possess modified gyrases or because their cell membranes are less permeable to the fluoroquinolones.

At about the same time that the quinolones were marketed, the much more complex streptogramins were introduced. These are also obtained from various *Streptomyces* species, and the drug combination of quinopristin and dalfopristin (Synercid) has good activity against Gram-positive bacteria including multiresistant strains. They appear to act on the peptidyl transferase domain of the 50S ribosomal subunit.

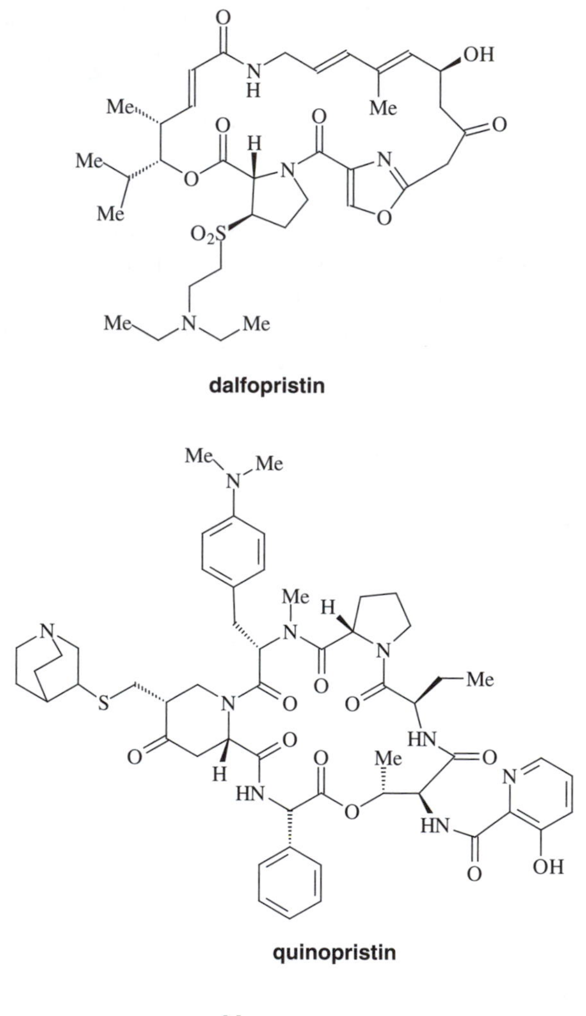

dalfopristin

quinopristin

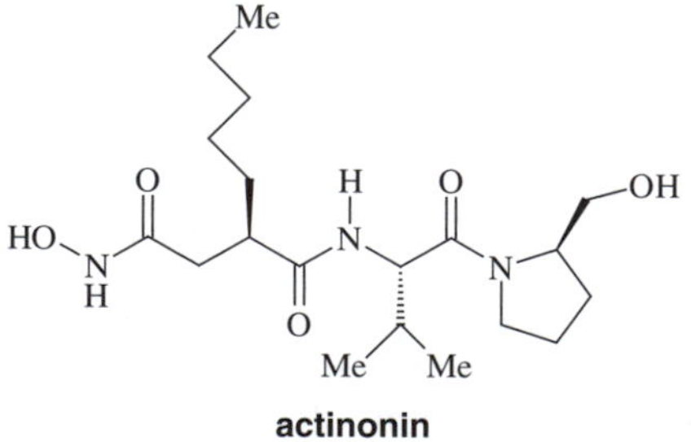

actinonin

It is disturbing to note that the only new class of antibacterial agents to appear during the past 30 years are the newly introduced oxazolidinones, of which linezolid and eperezolid are the first two examples – the former was approved in the USA (in 2000) and in the UK (Zyvox, in 2001). These emerged from work originally carried out by Dupont in the mid-1980s, which culminated in the discovery of the compound DUP 721. This had antibacterial activity but also possessed lethal toxicity in *in vivo* tests. Later

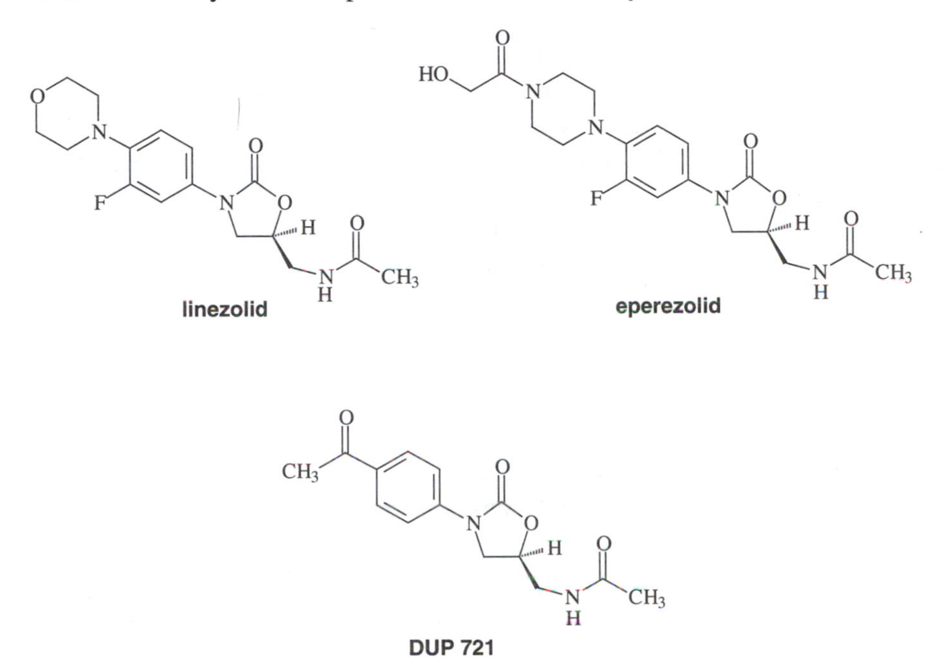

structure–activity work by Upjohn-Pharmacia (now part of Pfizer) demonstrated the requirements for effective antibacterial activity without associated toxicity, and the synthesis of these compounds in a stereochemically defined form was also highly efficient. As a bonus, it was discovered that the compounds exhibited a novel mode of action in that they bound to the 50S ribosomal sub-unit and inhibited the formation of the initiation complex that involves association of the 30S and 50S sub-units and various initiation factors. They thus inhibit protein biosynthesis before it can get started. Zyvox is orally active and is almost 100% bioavailable, that is, all the oral dose passes out of the gastrointestinal tract into the bloodstream. In tests against MRSA and vancomycin-resistant enterococci, the clinical responses for Zyvox range from 65 to 90%. However, only the future will tell whether the bacteria will soon develop resistance to these new drugs.

Another recent development also involves the initiation of protein biosynthesis. The first transfer RNA that binds to the bacterial ribosome (but not its human counterpart) is the one that carries a formylated methione, and there are key bacterial enzymes for the deformylation of this amino acid (the *N*-deformylases) at the end of polypeptide biosynthesis. One natural product – actinonin from *Streptomyces roseopallidus* and other species – is a potent inhibitor of these bacterial enzymes, and may offer a novel mode of antibacterial action.

In this new millennium, we can recognise three distinct periods in the last century: the first 35 years during which the threat from bacterial infection was essentially as it had been throughout the previous centuries; a golden age of antibacterial drugs of 25 years or so, during which the bacteria seemed upon the point of imminent defeat by these new wonder drugs; and the final 40 years, during which the pharmaceutical industry waged a ceaseless battle against an ever-burgeoning catalogue of resistant bacterial strains. The drug companies have produced literally thousands of novel semi-synthetic antibacterial drugs, but the bacteria have thus far always been able to respond. A return to the pre-antibiotic 'dark ages' is a real possibility. The ingenuity of chemists and biologists has never been of greater importance, and their efforts to design agents with new modes of antibacterial activity are proceeding with an unprecedented sense of urgency. The new agents will come, as so often in the past, from screening programmes with natural products, from synthetic compound libraries, but increasingly from a systematic study of the bacterial genomes and the knowledge of the enzymes that arise from these genes.

The task is, however, made no easier by the sudden appearance of totally new and terrifying virus-induced diseases like AIDS and Ebola haemorrhagic fever. The quest for 'magic bullets' to control these and other viruses is the subject of the next chapter.

Chapter 3

Antiviral Treatments

INTRODUCTION

Jenner's 'speckled monster' (smallpox) has been defeated, but AIDS will be with us for many years to come. While smallpox was eradicated by means of a worldwide vaccination campaign, prevention of acquired immune deficiency syndrome or AIDS will require new drugs and more careful sexual behaviour. Both diseases do share a common feature – they are caused by Nature's most successful 'parasites' – the viruses. In the developed world, it is not uncommon for a person to survive to a ripe old age without experiencing a serious bacterial infection or contracting one of the many forms of cancer. They will, however, have suffered from the effects of numerous viral infections of the respiratory tract, i.e., colds and flu, and most probably, from the common childhood virus-inflicted disease of chicken pox. It is unlikely that any of these afflictions will have been life-threatening, but they will have caused many days to be lost from school or work. In other words, the morbidity due to the common viral diseases is high, but the mortality is low.

In contrast, in the less affluent countries of Latin America, Asia and Africa, the measles virus alone claims the lives of many millions of persons each year. This should be contrasted with the two million deaths each year from malaria. The number of virus-related deaths in the developing world would be even larger but for the eradication of smallpox and the gradual introduction of immunisation programmes for the common childhood diseases like measles, whooping cough and polio.

Such prophylaxis has been almost universally available in the developed countries for the past quarter of the century, and the two most common reasons for seeking medical attention and therapy are respiratory tract infections and herpes infections, for which there are no effective vaccines. Around 35% of all acute medical illness can be ascribed to the 'common cold' and a further 40% to influenza. In addition, the sexual freedom of the past 40 years has resulted in up to 10% of the population being affected by recurrent genital infections caused by the herpes simplex virus. Fortunately, the human immunodeficiency virus (HIV) is less infectious than most other viruses, and the AIDS epidemic has not been as inexorable as it might have been if HIV was spread by droplet infection as with colds and flu. The impact of AIDS has, nonetheless, reminded mankind of their precarious place in a world full of dangerous pathogens.

WHAT IS A VIRUS?

Before we can begin to design methods of destroying viruses, it is important to know something of their constitution and lifecycle. Jenner was probably the first to use the term 'virus' to describe the infective agent present in smallpox or cowpox, and he took the term from the Latin 'slimy liquid or poison.' Pasteur also used the term virus to describe the causative agent of rabies, which he knew could not be a bacterium like the others he had studied. But it was Iwanowsky and Beijerinck who, between 1894 and 1898, provided the first definitive proof that the organism that caused tobacco mosaic disease could be separated from co-occurring bacteria, and appeared to be a new type of organism. Leoffler and Frosch (1898) demonstrated that a similar, non-bacterial organism was responsible for foot and mouth disease in animals, and finally in 1935, Wendell Stanley obtained a sample of the tobacco mosaic virus in crystalline form. This was still infectious and could reproduce within host cells. The advent of the electron microscope finally enabled biologists to classify the various virus types according to structure.

All viruses contain a discrete amount of DNA or RNA (their *genome*) either as a single strand or a pair of strands, together with a few non-structural proteins (enzymes) packaged within a protein coat called the *capsid*. In some viruses, this unit is in turn enclosed within a lipoprotein outer coat. These particles function as a living organism in the sense that they have the capability to reproduce, but unlike the bacteria, they cannot do this without eliciting the assistance of the cells that they invade. They are thus parasitic on their host cell.

The complete, mature virus particle is known as a *virion* and usually has a regular shape. Many virions are icosahedral, that is, the capsid is formed from identical protein subunits (*capsomeres*) that combine to produce a solid with twenty faces, each of which is an equilateral triangle. The herpes viruses are of this type, as are the picornaviruses of which the polio viruses and rhinoviruses (cold viruses) are the bestknown members. The other common regular shape is that of a helix, and the tobacco mosaic virus is of this type. Its single helical strand of RNA is enclosed within a hollow tube, which comprises 2130 protein subunits arranged in a helix. Other viruses with a similar structure are the orthomyxoviruses (*e.g.,* influenza), paramyxoviruses (*e.g.,* mumps), and the rhabdoviruses (*e.g.,* rabies). Finally, viruses may possess less regular shapes and the large pox viruses (*e.g.,* smallpox), which can attain sizes of 250 nm diameter, comprise DNA within a protein case, itself surrounded by a lipid bilayer with glycoproteins enbedded within it. The measles virus has a similar outer envelope but with RNA encased within a TMV-like helical protein.

A brief comment about size is in order, since viruses range from the minute, like polio viruses, rhinoviruses and hepatitis A (all around 25 nm), to the comparatively massive pox viruses, which are 10 times larger and just about visible under the light microscope. Whatever their size and shape, the one common feature of all viruses is their use of host-cell biochemistry to help with their reproductive cycle – their *replication*. It is usual to identify six stages in the life cycle of viruses:

- attachment to the host cell;
- entry into the host cell (internalisation);
- uncoating;
- replication of the genome to produce viral DNA and RNA, and subsequent production of viral proteins and enzymes;
- assembly and maturation of virus particles; and
- release from the host cell.

An understanding of the biochemical basis of these various processes should facilitate the design of drugs that interfere with them.

Attachment of the virus to its host cell requires a matching of a glycoprotein or lipoprotein on the viral surface with a complementary chemical species on the surface of the cell. This is usually termed *cell recognition* and is both specific and involves tight binding. It is another example of the 'lock and key' principle mentioned in Chapter One. For example, on the surface of HIV, there is a glycoprotein called gp120 and this is recognised by a matching glycoprotein CD4 on the surface of the T4 lymphocyte. Clearly, disruption of this interaction would provide a possible strategy for chemotherapy, and certain agents like the anticoagulant heparin, do act in this way.

Once recognition and subsequent tight binding have occurred, the virus must penetrate the cell membrane of the host cell. Those viruses that have an outer envelope become engulfed by the cell membrane in a process called *pinocytosis*. The viral envelope becomes fused with the cell membrane and ultimately this breaks open to release the virion into the cytoplasm of the host cell. Its coat protein is then digested and the genome (DNA or RNA) is revealed. Viruses that do not possess an outer envelope also associate with the host cell membrane but in this case a pore opens up and the viral particle passes through it into the cytoplasm (*endocytosis*), where it is uncoated.

Once inside the cell, the viral genome can elicit the help of the host cell's replication and protein biosynthesis capabilities in order to produce new copies of viral DNA or RNA, together with new structural proteins and essential enzymes. For the rhinoviruses (common cold viruses), their RNA is already able to function as a blueprint (messenger RNA, m-RNA) for protein production – they are termed RNA-(+)-viruses. The other major RNA viruses (*e.g.,* influenza) have to make a complementary copy of their RNA before this can function as m-RNA – these are termed RNA-(−)-viruses. Whether the virus is (+) or (−), its m-RNA codes for an RNA polymerase, which controls the production of new copies of viral RNA. To ensure that the correct (+) or (−) form is made, the RNA polymerase first catalyses the production of a complimentary RNA strand – (+) requires a (−) strand and (−) requires a (+) strand, and then uses this as a template for the production of the complimentary (native) RNA form.

Most of the viruses that contain DNA (*e.g.,* herpes) use this as a direct template for m-RNA production using host cell DNA-directed RNA-polymerase. However, the pox viruses have their own RNA polymerase. Finally, the so-called *retroviruses* like HIV use their RNA as a blueprint for the production of a complementary DNA strand with the help of an enzyme called *reverse transcriptase* – an RNA-dependent DNA polymerase. This nomenclature arises from the fact that these organisms operate in reverse of the normal cellular replicative process whereby DNA acts as a blueprint for RNA production. The DNA thus produced is incorporated into host cell DNA with the help of another viral enzyme – *HIV integrase* – and the cell ultimately produces new viral genomic RNA and viral proteins using its normal replication procedures.

The requirement for the intact and functioning host cell mechanisms means that an ideal anti-viral drug must somehow damage the virus without harming the host cell. This clearly represents a major design problem, but there are a few windows of opportunities. The RNA viruses use RNA-directed RNA polymerases to make copies of their genome, and these enzymes do not occur in normal animal or plant cells. Similarly, the reverse transcriptases of the retroviruses are also unique to these species, and provide viable targets for chemotherapy. Finding specific inhibitors for the DNA-directed RNA polymerases of the DNA viruses is obviously more difficult, because these are similar to the enzymes of animal and plant cells. However, there are always subtle differences in structure and biological specificity, and these have been exploited in the therapy of, for example, herpes infections. In addition, certain viral enzymes with key roles in the construction of the subunit structures of RNA and DNA are slightly different from the equivalent host enzymes, and a number of drugs can exploit these differences.

The final stages of the viral replication involve the assembly of the new viral RNA or DNA and viral enzymes within the mature virus capsid, and its subsequent expulsion from the host cell. The proteins that form the capsid have the remarkable capacity of self-assembly, and once a sufficient number of these proteins have been produced, they form the viral coat enclosing the RNA/DNA and replicative enzymes. Eventually, the host cell's store of energy and materials is exhausted, and it dies, thus releasing thousands of mature virions to infect other cells. This simple scenario holds for those viruses that do not possess an outer envelope. The more complex enveloped viruses are released from the cell by a process know as *budding*. Initially, their coat protein becomes associated with the host cell membrane and other enzymes control the inclusion of RNA/DNA and replicative enzymes within a new capsid. Once this has formed, the virus envelope develops out of the host cell membrane and effectively seals the virus from the cytoplasm. Finally, a swelling or protuberance appears on the exterior of the host cell and the mature viral particle is released. Cell death is not necessary for this release, although the cell almost inevitably dies due to exhaustion of its store of energy material. These various maturation processes clearly present further opportunities for drug intervention, especially since they are essentially unique to the viruses, and one of the newest drugs for the treatment of influenza acts at this stage of the replication.

The rest of this chapter will show how the most successful treatment modalities have developed. These include: vaccination (for smallpox, polio, measles, *etc.*), and chemotherapy using anti-viral drugs. A large number of viral diseases still lack an effective means of treatment, and the chapter will also cover the attempts to treat the common cold and influenza; the struggles with HIV; and the emergence of viruses (Marburg, Ebola, Lassa) that cause haemorrhagic fever.

ANTIBODIES: NATURE'S MAGIC BULLETS

Smallpox

It is said that the invading Hun never washed and wore their clothes until they rotted. In addition to the more obvious affects of their rape and pillage during the period 370–451 AD, they are reputed to have brought smallpox to Europe from their Asian homelands. However, a Roman legion is also credited with bringing the so-called Antonine plague (but probably smallpox) back from a campaign in the Middle East in 165 AD, and this subsequently killed about one-quarter of the population of their empire (possibly as many as five million people). Galen, the famous physician of the age, accurately described the fever and the 'many lesions which changed into ulcers' that are typical of smallpox. Whoever was responsible for its introduction into Europe, there is no doubt that *poc* or *pockes* (the Old English for a bag or pouch) caused periodic epidemics throughout Europe, but was never as feared or devastating as bubonic plague, also introduced from the Far East. It caused more obvious loss of life in the Caribbean, and Central and South America during the 16th century, when over half of the estimated population of 50-million perished from smallpox, influenza, and measles introduced by the European adventurers. The devastation was no less serious in the evolving colonies in North America. Writing in 1634, John Winthrop, Governor of Massachusetts claimed:

> they (the Natives) are near all dead of smallpox...the Lord hath cleared our title to what we possess.

An early example of biological warfare is also alleged to have occurred when Sir Jeffrey Amherst, who was commander of the English forces in North America, ordered that smallpox-contaminated blankets be distributed to the local native people. The effects on a population that had never been exposed to this virus were devastating.

The situation in Europe changed towards the end of the 17th century. Improvements in sanitation produced a dramatic decline in the incidence of bubonic plague, whilst a mutation of the variola virus (the causative agent of smallpox) appears to have produced a more virulent strain. During the 18th century, most Europeans contracted smallpox and around one in five of the population (a total in excess of 50 million people) died from the disease. Localised epidemics could be even more devastating like the one of 1707 in Iceland, when 36% of the population died, or the one of 1719 in Paris, when 14000 of the population died.

The disease was originally simply called 'the pox' but became smallpox following the introduction of 'the Great Pox' or syphilis at the end of the 15th century. Two variants are now known to have existed – *Variola major*, which caused a high mortality rate, and *Variola minor*, which was responsible for around 1% mortality. Like all the most successful viruses, it was spread by droplet infection and even inhalation of air in the vicinity of a victim's body could produce infection. The initial symptoms were typical of the majority of viruses – headache, general malaise and fever – and these developed around nine days after the initial infection. Two to three days later, the rash appeared, first on the face and then this spread to cover the whole body. The small red spots developed into pustules, which, if ruptured, oozed a malodorous pus that was highly contagious. Patients often appeared bloated by the almost continuous expanse of pustules and many died from internal haemorrhage. Unbroken pustules formed a scab that lasted for about three weeks, and survivors often required many more weeks of careful nursing if they were not to succumb to post-viral infection. Many were left horribly scarred and in some cases blind. As Thomas Macauly wrote in his *History of England*:

> Smallpox is the most terrible of all the ministers of death…leaving on those whose lives it spared the hideous traces of its power, turning the babe into a changeling at which the mother shuddered, and making the eyes and cheeks of the betrothed maiden objects of horror to the lover.

Such was the fate that befell Lady Mary Wortley Montagu, a woman of considerable beauty whose face was scarred by the effects of smallpox in 1715. In 1716, she accompanied her husband when he took up the post of Ambassador to Turkey in Istanbul. Here, she took a close interest in the Turkish method of inoculation that was used as a preventative measure against smallpox. In her now justly famous letter to her friend Mary Chiswell in 1717, she gave a graphic description of the technique:

> The smallpox, so fatal and so general amongst us, is here entirely harmless by the intervention of ingrafting, which is the term they give it. There is a set of old women who make it their business to perform the operation, in the month of September, when the great heat is abated… . The old woman comes with a nutshell full of the matter of the best sort of smallpox, and asks what veins you please to have opened. She immediately rips open that you offer to her with a large needle (which gives you no more pain than a common scratch), and puts into the vein as much venom as can lie upon the head of her needle…and in this manner opens up four or five veins.
>
> The children or young patients play together all the rest of the day, and are in perfect health to the eighth. Then the fever begins to seize them, and they keep their beds two days, very seldom three. They have very rarely above twenty or thirty (pustules) in their faces, which never mark; and in eight days' time they are as well as before the illness… . Every year thousands undergo this operation; and the French ambassador says pleasantly, that they will take the smallpox here by way of diversion, as they take the waters in other countries. There is no example of anyone that has died in it; and you may believe I am well satisfied of the safety of the experiments, since I intend to try it on my dear little son.

The practices that she observed were not new and had been used in Turkey, Greece, China and India for several centuries. She was also not the first English person to describe the technique, since Dr. Clopton Havers read a paper about the technique to the Royal Society in 1701. This august body then discussed the matter in both 1713 and 1714, but the medical fraternity was not yet ready to adopt these foreign methods.

Before returning to England, Mary Wortley did have her son inoculated, eliciting the help of Charles Maitland, the surgeon to the British Embassy in

Istanbul. He described the outcome in a small book entitled *Account of Inoculating the Small Pox* published in 1723:

> On the third day, bright red spots appeared in his face, then disappeared…till in the night betwixt the seventh and eighth day, he was observed to be a little hot and thirsty…then the smallpox came out fair… . He had above one hundred (pustules) in all upon his body, but without any the least disorder…they all fell off, without leaving any one mark or impression behind them.

In 1721, there was a particularly bad outbreak of smallpox in London, and Mary Wortley asked Maitland to inoculate her younger daughter. This is believed to be the first such inoculation in England, although similar techniques were already in use in both Wales and Scotland. Maitland was obviously encouraged by his successes since he persuaded the College of Physicians to allow him to carry out a highly unethical 'clinical trial' in 1721. Six condemned men in Newgate Prison were given the option of a reprieve if they would undergo inoculation. All accepted and were operated on without problems.

At about the same time, Dr. Zabdiel Cutler, a Boston physician, carried out probably the first inoculation in America when he treated his son and two negro slaves with a 'sharp toothpick and quill'. All suffered mild infections and became immune to smallpox. Cutler had been encouraged to carry out these inoculations by the controversial cleric Cotton Mather, and once news of the operation became public knowledge, Cutler had to hide from what was essentially a lynch-mob. This was hardly surprising given that the Puritans of New England believed that all illness was retribution from God and was to be borne with stoicism. Despite these experiences, Cutler went on to write *An Historical Account of Smallpox Innoculation in New England*, and became well-known for his studies in England, with election to membership of the Royal Society in 1726.

The well-to-do inhabitants of London were quite taken by this new medical discovery and a growing number, including the Prince of Wales, had both their staff and children inoculated. The practice might have become even more popular but for two adverse reactions: firstly, a few, hitherto healthy patients died as a result of their inoculations, and secondly, the clergy began to rail against this "dangerous and sinful practice." That some deaths would occur was almost inevitable because somewhere along the line, the inoculation technique had changed from the simple scarification and introduction of minute amounts of virus, to one that involved what one physician described as a "long and large incision" with probable incorporation of large amounts of virus. In addition, what was often forgotten (or even not realised) was that although those inoculated usually suffered only a mild attack of smallpox, they were nonetheless infectious and carriers of large numbers of virus particles.

Clergymen, especially in the provinces, warned their parishioners of the 'heathenish and diabolical practice' and warned of divine retribution. The medical profession also advised caution; and against this background of professional disquiet and clerical opposition, Lady Mary Wortley carried out a virtually one-woman campaign enthusiastically exhorting people to have their children inoculated. It is interesting to compare these concerns with the periodic panics that are associated with the various modern childhood vaccinations. The recent fears that the measles, mumps and rubella triple immunisation (MMR) is associated with a low incidence of autism and/or irritable bowel disease, is a case in point. There is always a slight risk associated with immunisations, but the benefits invariably far outweigh the infinitesimal risks.

During the period from 1721 to 1728, records kept by Guys Hospital in London indicate that the number of people receiving inoculation ranged from 469 (in the period 1721 to 1723) to 37 in 1728, and by the end of the next decade, the practice had all but ceased. In contrast, in the American colonies, inoculation was widely practised, especially during the epidemics of 1736–1737 in Philadelphia and of 1738 in Charleston, South Carolina. And it was an American physician, James Kilpatrick (later Kirkpatrick), who did much to revive the enthusiasm for inoculation in England. He founded the Smallpox and Inoculation Hospital in London in 1746, a year when there was a serious epidemic in the city resulting in 3236 recorded deaths. Other hospitals were soon established, especially to treat children and the poor, and special inoculation houses were set up outside the capital.

This slowly growing acceptance of inoculation changed to an almost universal enthusiasm in the early 1750s, primarily due to the devastating effects of the nation-wide epidemic of 1751–1753, but also because of the new, safer technique introduced by the Suffolk surgeon, Robert Sutton. The epidemic killed more than three-and-a-half thousand in London and was ultimately responsible for many tens of thousands of deaths. The College of Physicians dropped their resistance to inoculation and issued a report that declared the practice to be 'highly salutary to the human race'; and even the clergy dropped their opposition, preferring to accept that God had showed man how to overcome the affliction.

Robert Sutton's main contribution was to rediscover the Turkish method of minor scarification with introduction of small amounts of 'unripe' material from immature pustules. This ensured the induction of immunity without giving rise to a serious attack of smallpox. Initially, much of the inoculating was carried out in special residential centres with patients staying as long as a month, and receiving special diets and exercise programmes, at a cost of around 20 guineas. This charge fell rapidly during the 1760s as more physicians, and subsequently apothecaries or even lay-persons, sought participation in this lucrative business. Fees as low as five guineas for the gentry and three for their servants soon became the norm, but the most successful 'inoculators' could still become very wealthy. Daniel Sutton, who took over his father's business during the 1760s, is claimed to have inoculated (with several assistants) more than 20,000 persons during the period 1764–1766, and eventually owned inoculation houses throughout southern England.

These charges were beyond the means of the working class; hence, when an epidemic occurred in a town, mass inoculation was undertaken, with the parish footing the bill for those who could not pay. The knowledge that last-minute inoculation could prevent infection even when a new epidemic had been declared at last took away some of the fear associated with smallpox.

Several physicians went as far as to suggest the possibility of eradicating the disease through a nation-wide campaign of inoculations, but this was never pursued seriously. There is, however, no question that the greatly reduced fear of the disease and the higher probability of children reaching adulthood (probably about 15% greater) contributed to the increase in population in England and Wales of six to nine million during the second half of the 18th century.

If Lady Mary Wortley's claim to fame was her introduction of a practice that helped control epidemics of smallpox, then Edward Jenner secured his place in history by discovering a method that would ultimately lead to the eradication of the disease. Despite the popularity of inoculation in England and Wales – the Suttons and their assistants probably inoculated 400,000 people during the latter half of the 18th century – the practice was less popular in Europe. The French overcame their initial reluctance to adopt the practice following the death of Louis XV from smallpox in 1774; and Catherine the Great popularized inoculation in Russia by having her own family inoculated. But physicians in both Austria and Germany were generally antagonistic, being justly worried about the possibilities of cross-infection when using a live virus. Pockets of the disease were thus always present and epidemics could arise at any time. Eradication was impossible.

The means to achieve this apparently unattainable goal had their origins in the centuries-old observation that those who were actively involved with dairy cows rarely contracted smallpox. Jenner's seminal contribution was to establish that this was linked to their contraction of cowpox, a relatively minor variola-induced disease that caused cows to become moderately unwell and produced eruptions on their udders. Milkmaids and others who milked infected cows usually became unwell with a slight fever and (in Jenner's own words):

> ..the pulse is quickened, and shiverings with general lassitude and pains about the loins and limbs come on. These symptoms, varying in their

> degree of violence, generally continue from one day to three or four, leaving ulcerated sores about the hands.

Jenner, an unassuming Gloucestershire physician, was also an expert inoculator, and his suspicions about the protective affect of this infection first arose in 1778 during a local smallpox epidemic. In his subsequent book of 1801 entitled *The Origin of the Vaccine Inoculation*, he told of his discoveries:

> My attention to this singular disease (cowpox) was first excited by observing, that among those whom in the country I was called upon to inoculate, many resisted every effort to give them the smallpox. These patients I found had undergone a disease they called the cowpox.

But this was not the first description of the possible association since an experienced inoculator by the name of Dr. Fewster had sent a report to the Medical Society of London in 1765 entitled *Cowpox and its ability to prevent smallpox*. It is also alleged that a Dorestshire farmer, Benjamin Jesty, inoculated his wife and two sons using cowpox virus taken from an infected udder. Jenner's own experimental contribution was much more scientific, if wholly unethical by today's standards. In a letter of July 1796 sent to his friend, the wine merchant Edward Gardner, he described his first experiments carried out in May 1796:

> I have at length accomplished what I have been so long waiting for, the passing of the vaccine virus from one human being to another by the ordinary mode of inoculation. A boy of the name Phipps (aged eight years) was inoculated in the arm from a pustule on the hand of a young woman (Sarah Nelmes) who was infected by her master's cows. Having never seen the disease but…when communicated from cow to the hand of the milker, I was astonished at the close resemblence of the pustules in some of their stages, to the variolous pustules. But now listen to the most delightful part of my story. The boy has since been inoculated for the smallpox which, as I ventured to predict, produced no effect.

Encouraged by this result, Jenner proceeded to inoculate more than a dozen other children all of whom developed cowpox and then became donors of virus to other recipients. Most of these children were then challenged with a smallpox inoculation, yet, none of them succumbed to infection. This was just as well because such unethical behaviour would surely have resulted in his arrest and imprisonment.

The significance of Jenner's experiments was quickly appreciated by physicians in London, and he was encouraged to publish his findings in 1798 in a report entitled *An inquiry into causes and effects of the variolae vaccinae*. Jenner invented this name for the cowpox virus on the assumption that it was related to smallpox (*Variola major*) but came from the cow (Latin *vacca*). The term *vaccination* is said to have been first employed by a Plymouth surgeon, Dunning, who used it to describe an inoculation with cowpox. In his publication, Jenner pointed out that inoculation with cowpox had two important advantages over the corresponding operation with smallpox. Firstly, it was cheaper and easier to perform, and produced a much milder form of infection. Secondly, and most importantly, there was no chance of cross-infection since cowpox was not transmitted by droplet infection. This was a serious problem with smallpox inoculation and probably resulted in at least one death per hundred persons inoculated.

Jenner's fame spread and in 1800 he was asked to vaccinate the troops commanded by the Duke of York, prior to their departure for the campaign against Napoleon. The latter was keen to learn about the technique and dispatched one of his physicians, a Dr. Aubert, to England to gather information. In due course, he published his findings in a report entitled *Rapport sur la Vaccine* in 1801, and vaccination institutes were soon established all over France. News of the technique spread throughout Europe and as far as India and Iraq.

Recognition of Jenner's achievements came in the form the of a grant from Parliament of £10,000 in 1802, and by the formation of the Royal Jennerian Society in 1803 with the Queen as the patron and Jenner as its first president. The Society set up 13 vaccination stations and carried out more than 12,000 vaccinations in its first year of existence.

However, dissenting voices began to make themselves heard. There were those who were disgusted by "the introduction of a bestial disease into the human body," but more importantly, physicians drew attention to the impermanence of the immunity against smallpox conferred by vaccination. The Gillray cartoon of June 1802 with its depiction of mass vaccination and apparent problems with beast-related side-effects, nicely conveys the mood of the time. There is no doubt that many physicians were disturbed by the simplicity and cheapness of the technique, and by their loss of earnings as people turned away from smallpox inoculation. George Lipscomb, surgeon at St. Bartholomew's Hospital in London, published a pamphlet in 1805 in which he warned:

> …the inoculated cowpox is sometimes a severe and dangerous disease, and sometimes fatal…that it is productive of many horrid and loathsome symptoms and that vaccination ought to be immediately, and for ever, abandoned.

Such condemnations notwithstanding, the London Vaccine Institution opened in 1807 and by early 1808 had vaccinated 52,000 people free of charge.

During the next 30 years, vaccination became widely accepted and inoculation was left increasingly in the hands of amateurs and charlatans.

Unfortunately, the nation-wide smallpox epidemics of 1816–1819 and 1825–1826 revealed the serious limitations of vaccination, primarily with regard to the impermeance of the immunity, and inoculation became popular once again. The terrible epidemic of 1837–1840, during which 42,000 people died in England and Wales, reinforced the need for some kind of national campaign of vaccination in order to eradicate the disease. All the time inoculation was still available there was a reservoir of the virus in the community, and in order to overcome this problem, the Parliament passed a bill banning the practice in 1840. A second bill in 1853 made vaccination compulsory for every child before they were three months old, and Acts of 1861, 1867 and 1871 introduced tough penalties for non-compliance.

Inevitably, this encouraged a growing band of anti-vaccinators who opposed this infringement of civil liberties. Local vaccination officers were appointed to enforce the laws and the number of prosecutions rose steeply. By 1873, vaccination had become an election issue and a number of strongly worded publications appeared during the run-up to the election of 1874. For example, the Reverend George Cardew, rector of Helmingham, Suffolk was vehemently opposed to vaccination and wrote a pamphlet entitled *Think before you vaccinate* in which he stated:

Edward Jenner and Inoculation: A Gillray cartoon (1802) (From the Wellcome Trust Medical Photographic Library)

"... (vaccination involves) the cutting with a sharp instrument of holes in your dear little healthy babe's arm...and putting into the holes some filthy matter from a cow – which matter has generally passed through the arm of another child. So that your babe, just after God has given it to you, is made to be ill with a mixture of curruption of both man and beast."

The towns of Leicester and Banbury became key centres of public foment and a demonstration in Leicester in 1885 attracted 20,000 participants, not too surprising given that 4000 local people were awaiting prosecution for failing to have their children vaccinated.

The anti-vaccination stance of the public hardened, and in 1898, the Parliament bowed to this pressure and passed an act that allowed for the possibility of conscientious objection on the grounds that 'vaccination would be prejudicial to the health of a child'. This was extended in 1908, making it easier for parents to refuse vaccination; thus in a sense, the anti-vaccinators had won their case. However, this section of the population was very much in the minority, and general acquiescence had by now all but eradicated endemic smallpox from Britain. In 1908, for example, the only reported cases were among persons who had imported the disease from abroad, mainly via the busy seaports of Southampton and Liverpool. These infected individuals were quarantined in isolation hospitals including a floating hospital on the Thames near Dartford, and as a result, *Variola major* disappeared from the general population. A new variant, *Variola minor*, now became more significant. It appeared first in Southern Africa and the West Indies, then spread to Brazil, North America and Europe. Since mortality from this disease was very low (less than 1%) and vaccination provided full protection, this caused little concern. Eventually, growing concerns about the slight risk of neurological complications following vaccination led to the abolition of routine infant vaccination in 1974. The 'speckled monster' had been defeated, at least in Britain and most of Europe.

The situation in the 1950s was much less rosy for the less developed parts of the world, with an annual death toll of around two million from the disease and around 10 to 15 million cases world-wide. In 1958, the Soviet Union persuaded the World Health Organisation to mount a global campaign of eradication, and this commenced in 1967, with an avowed aim of providing at least 250 million vaccinations per year. Hundreds of dedicated WHO field-workers scoured the world's most obscure and often dangerous places, seeking isolated pockets of infection. They were equipped with a new bifurcated needle that when dipped into the vaccine would pick up exactly the right dose, and could be sterilised by boiling water ready for reuse. There were several serious epidemics in the early 1970's – one killed 25,000 persons in NE India in 1974 – but by 1975, they had cornered the last remaining carriers of the virus in Bangladesh. Despite the coincident civil war and devastating floods, they managed to identify and treat the last victim of *Variola major,* a Bangladeshi girl called Rahima Banu, in November 1975. A few cases of *Variola minor* were reported in Somalia in 1977 during the war between Somalia and Ethiopia over the disputed Ogaden region. One other fatal case of smallpox was reported in 1978, when a laboratory assistant at the University of Birmingham, England, was infected by an illegally held sample of the virus. Ultimately, on May 8th, 1980, the WHO formally declared that smallpox had been eradicated from all parts of the world. This was not quite true, since one laboratory in the USA, the US Center for Disease Control in Atlanta, and the corresponding facility in Moscow, maintain samples of the viruses, although these are due to be destroyed once gene sequencing studies have been completed. This is likely to coincide with the bicentenary of Jenner's first experiments with cowpox (now known simply as vaccinia) and will make a fitting tribute to his seminal contributions to virology. His fame was initially marked by the erection of a statue in Trafalgar Square, London, in 1858. However, his presence in an area dominated by famous soldiers and sailors was deemed to be inappropriate in 1862, and his statue was banished to Kensington Gardens. The satirical magazine *Punch* noted:

> England's ingratitude still blots
> The escutcheon of the brave and free;
> I saved you many million spots,
> And now you grudge one spot for me.

Rabies

Historically, the discoveries made by Pasteur in his work on anthrax and rabies provide the link between the relatively crude technology of Jenner and the sophisticated vaccines of today. During the 1870's, a vigorous debate had been conducted by the Paris Academy of Medicine with subject to the relationship between smallpox and cowpox, and Pasteur would have been aware of the prevailing view that cowpox was a form of smallpox that had been attenuated by passage through the cow. This had, in fact, been suggested by Jenner almost 100 years earlier. Pasteur's contribution to the debate arose from a chance discovery in his studies on the chicken cholera bacillus. During the summer of 1879, the cultures of this bacillus lost their potency, and new cultures were generated using samples of the bacillus from a fresh outbreak of the disease. When material from these new cultures was inoculated into chickens that had never encountered the bacillus, they were killed; but inoculation of chickens that had been treated earlier with the old, relatively weak culture, produced no deaths. Clearly, even a relatively non-potent bacillus could provide some immunity, and it seems that Pasteur perceived the similarities between this attenuated bacillus and attenuated smallpox in the form of cowpox or vaccinia.

His first successes with attenuated vaccines were with the anthrax bacillus, but it was his later work with rabies that revealed the enormous possibilities of attenuated vaccines. The word rabies is derived from the Latin *rabere* – to rave, and the causative virus is usually passed to humans by the bite of a rabid dog or a wild canine (fox, wolf, *etc.*), and very occasionally by a bat. Certainly, wild canines are the natural reservoirs of the virus. The incubation period is typically 20 to 90 days, and after this, someone unlucky enough to be bitten by a rabid animal will suffer a period of malaise, mild fever and headache for a few days, followed by a period of excitement, when the patient cannot sleep and becomes apprehensive and easily alarmed. There is intermittent fever and some localised paralysis, especially of the neck and throat muscles, which makes swallowing problematic and is probably responsible for the fear of water (hydrophobia). More severe paralysis follows and this leads inexorably to cardiovascular collapse, coma and death, all within about one week of the appearance of the first symptoms.

Although the incidence of rabies in France was not high at the time of Pasteur's investigations, it was a disease that was invariably fatal; hence, a life-saving therapy would have a major impact. In his early experiments, he took tissue from an infected dog and inoculated this into the brain of a rabbit. After the rabbit's death, he then used tissue from the brain and injected this into either rabbits or dogs. The rabbits died but the dogs survived, hence establishing that the virulence had increased for the rabbit but had been attenuated for the dog. But his optimum method of attenuation involved an excised spinal cord from an infected rabbit, which had been treated with potassium hydroxide to delay putrefaction, and then dried in an atmosphere of sterilised air for around two weeks. This produced a highly attenuated vaccine, although more virulent vaccines could be produced if shorter periods of drying were employed. These materials were used in life-saving vaccinations of dogs that had been bitten by rabid animals; but there was no certainty that it would work for humans.

Pasteur was understandably reticent about a human trial, but in July 1885, he was forced into taking this crucial step. An eight-year-old boy, Joseph Meister, with multiple bites inflicted by a rabid dog, was injected with dried spinal cord tissue. He received 12 successive injections of increasing potency and did not develop rabies or display any of its symptoms. Other successes followed and within 15 months, almost 2500 people had received the vaccine.

These days, the vaccine is produced using cultures of human cells in which the virus is grown and then attenuated, by heat treatment or with chemicals. Although there is little need for the vaccine in Europe, rabies is still a major problem in Russia, Africa, China, Japan, and the USA. In India, the disease claims the lives of fifteen to twenty thousand people every year, an unsurprising statistic when one notes that there are 50 million dogs in the country.

Like Jenner before him, Pasteur made a major if somewhat serendipitous contribution to modern immunology. His written comments in 1885 were highly prescient:

> I am inclined to believe that the causative virus of rabies may be accompanied by a substance which can impregnate the nervous system and render it thereby unsuitable for the growth of the virus. Hence rabies immunity.

He was, of course, unaware that the 'substance' was the glycoproteins on the surface of the virus, which excited the attention of the immune system and induced the production of antibodies to the virus. As more has been learnt about the surface structure of viruses, and how this elicits an immune response, the design of synthetic vaccines that are both safer and more effective has become possible. The evolution of the various vaccines against poliomyelitis provides a good example of these developments.

Polio

The poliovirus has probably coexisted with humans from ancient times. Its normal route of entry into the blood stream is via the gastrointestinal tract – it is an enteric pathogen – but it can also invade the nervous system and may then produce severe paralysis. The German physician Erb first used the term 'acute anterior poliomyelitis' in 1875, although the condition was also known as 'essential paralysis of childhood'. It seems that epidemics of infantile paralysis are a relatively recent phenomenon, and prior to the Industrial Revolution, the virus usually caused a mild flu-like condition that was resolved within a few days. The change in epidemiology may be due to a mutation in the virus, but is more likely due to improvements in living conditions. Since the virus can survive in the faeces for as long as three weeks, this used to be the primary mode of spread. Babies received minor immunity via breast milk, and this was sufficient to help them fight subsequent infections. As a consequence, it is likely that all adults had acquired immunity to the virus. Once sanitation was improved and people moved from the country to the towns and cities, exposure to the virus became less frequent, and babies were less likely to gain immunity from their mothers. An infection in infancy or early adulthood now became a more serious event.

The incubation period is typically two to three days and this is followed by a minor illness typical of any virus: malaise, slight fever and a mild sore throat. In acute infections, this is followed by what is known as the meningitic phase associated with headache, fever and vomiting, and this may resolve after a week or so. However, it may also precede the paralytic stage in which the virus invades the nervous system and destroys the neurons. This can lead to paralysis of all skeletal muscles with resultant inability to swallow, move and breathe. Recovery is possible, although most of these patients never recovered all of their mobility, and some of those afflicted in the epidemics of the 1940s and 1950s, spent the rest of their lives in an iron lung.

The first polio vaccine became available in 1955, thanks to the pioneering work of Jonas Salk and his collaborators at the School of Medicine of the University of Pittsburgh, and this became known as inactivated polio vaccine (IPV) since it involved treatment of virulent strains with formaldehyde. This vaccine had to be injected and although multiple doses provided good protection against infection, its relative lack of potency provided little immunity against wild poliovirus in the gut. Spread of the virus was thus still possible. This vaccine has been largely replaced by oral polio vaccine (OPV), which comprises attenuated live virus, and this induces excellent immunity when taken orally. The virus is grown in human or monkey cells and the use of this vaccine has led to the near eradication of endemic virus in the developed world. The situation is not so satisfactory in the developing countries. Global totals for reported cases have fallen from 25.7 thousand in 1988 to less than four thousand in 1994, and the World Health Organisation set itself the goal of global eradication by the year 2000. But this proved not to be so easily attainable as with smallpox. Those suffering from smallpox were clearly ill, and were never silent carriers of the disease, while those suffering from the mild form of polio excrete the virus continually for up to three weeks.

Polio is not the only disease that the WHO would like to eradicate. The virus-induced diseases like polio, measles, (viral) meningitis, and cholera, together with malaria, typhoid and pneumonia and other bacteria-induced conditions, currently cause the deaths of more than 13 million children under the age of five in the developing countries. Attenuated live vaccines against measles, mumps and rubella are widely available, as are those based on killed organisms of influenza and pertussis (whooping cough). But new, safer and more potent vaccines are desperately needed, and it is usually the case that vaccination is highly cost effective, especially in comparison with the use of drugs to treat overt disease. The advent of gene technology in the 1970s paved the way for new, synthetic vaccines, for at least some of these diseases.

Gene Technology

In its simplest form, the technology involves using a bacterial plasmid (see the section on antibiotic resistance) from *E. coli.* Its circular strand of DNA is cleaved with a nuclease enzyme (now called *restriction enzymes*), and then a DNA fragment that codes for the surface (glyco)protein of the virus is inserted using a DNA ligase to form a new circular DNA construct. The plasmid is then allowed to replicate in cells of *E.coli* or yeast to produce copies of the (glyco)protein for use as an immunological stimulant, that is, for vaccine production. Very often, it is useful to select those *E.coli* or yeast cells that contain the new plasmid; hence, the plasmid also carries a gene that confers resistance to a particular antibiotic. The cells are then grown in the presence of that antibiotic and only those that have the new plasmid can grow under these culture conditions.

Using this technology, it has been possible to prepare vaccines that are specific for the most immunostimulant (immunogenic) parts of the viral surface structure. For example, Albert Sabin was the first to show that there were three serotypes of the polio virus (PV1, PV2 and PV3), each of which had slightly different surface proteins (VP1, VP2 and VP3) and consequent subtle differences in immunogenicity. The gene sequences coding for these proteins were identified, and then inserted into *E.coli* and other carriers. Expression of these gene products allowed harvesting and purification of the discrete proteins, which could then be used for vaccine production.

Various hybrids of VP1 and hepatitis B surface antigen or the cell membrane proteins of *E.coli* have been prepared in this way, and have shown to be highly immunogenic. Similarly, hybrids of PV1/PV2 and PV1/PV3 have also been prepared. It is hoped that one or more of the vaccines prepared using these minimal forms of the virus will be more potent, even safer, and will have greater heat stability (an important factor in the developing world) than the existing vaccines.

This technology was also used to great effect for the production of *interferons*. Just as one bacterial infection can sometimes provide protection against another, through stimulation of the body's immune system, a similar phenomenon exists with viral infections. This viral interference was first observed by a number of research groups in the 1930s, but an explanation was only provided in 1957. Alick Isaacs and Jean Lindemann, of the National Institute for Medical Research in London, demonstrated that a protein, which they called interferon, was released by cells that had been infected with a virus. This agent then elicited an antiviral response by other cells.

It was quickly established that interferon was not single protein, but rather, a large family of proteins. These are now classified as α-interferon (more than 24 species) and β-interferons, which are produced by most types of cells when invaded by a virus, and γ-interferon, which is produced by infected T-lymphocytes. Once released, the interferons attach to specific receptors within other cells, and induce the synthesis of more than two dozen different kinds of proteins that provide resistance to viral attack. This is effected in a variety of ways that include inhibition of viral uptake and uncoating; inhibition of viral mRNA production and thence of viral proteins; and also inhibition of assembly of new virions and their release from the host cell. In short, disruption of just about every part of viral replication. Interferons also interact with other factors involved in the immune system, and they can potentiate the antiviral effects of these factors.

Not surprisingly, as the biological effects of the interferons were discovered, they became attractive targets for commercialization. Initially, they were produced by a process that involved incubation of human leucocytes with an inducer virus, and then by carrying out a laborious purification of the proteins. But the amounts available by this route were very small. The techniques of genetic engineering were then used to overcome this problem of supply. Leucocyte mRNA, specific for interferon, was used as a blueprint for the production of the complementary DNA, and then this was spliced into a bacterial plasmid DNA. This provided a whole 'library' of recombinant DNAs. From this 'library', it was possible to identify separate plasmids that had incorporated DNA sequences coding for the interferons. These specific DNA fragments were then isolated and spliced into the DNA of *E. coli*, and the organism could then be grown in culture with production of large amounts of interferons.

The availability of reasonable quantities of pure interferons allowed clinical trials to begin. In the mid-1980s, there was considerable excitement about the possibilities for antiviral and anticancer therapies, but the initial clinical promise of these drugs has not been realised. The α-interferons are currently licenced for the treatment of chronic infections caused by hepatitis B, C and D, and for the treatment of persistent genital warts (papilloma virus), although the dosage levels are quite high. This class of interferons also has some value for the treatment of hairy cell leukaemia (so-called due to the appearance of the malignant cells), and for Kaposi's sarcoma, which is a common malignancy affecting patients with AIDS. There was also a flurry of excitement, a few years ago, when the β-interferons were shown to have some benefit in the treatment of multiple sclerosis. The long-term benefits are yet to be established, but several forms are licenced for clinical use in multiple sclerosis.

At present, one exciting development in vaccine technology involves the use of *engineered living carriers*. These are viruses such as herpes simplex (cold sore virus) and vaccinia (cowpox), or avirulent strains of bacteria like *Salmonella typhimurium* and *E. coli*, into which the gene coding for the immunogenic protein of the pathogenic virus is inserted. Since these carrier organisms are relatively harmless to the person receiving them, they provide a safe vehicle for smuggling in portions of the more pathogenic virus, which will elicit the desired immune response. Some success has already been achieved with gene inserts from rabies, hepatitis B and even HIV, and it is remarkable that 200 years after Jenner discovered the utility of vaccinia in the prevention of smallpox, the same organism is the virus of choice for these new gene-based technologies.

COUGHS AND COLDS AND INFLUENZA

Influenza:the disease

November 1918 will always be remembered for two momentous events: the Armistice that ended World War I and the start of the great flu pandemic. The Great War claimed the lives of around 10 million people during a four-year period; the flu pandemic caused more than 20 million deaths in less than two years. The first cases of the flu were originally thought to have occurred amongst army recruits in the USA in the spring (there were 48 deaths). However, there had been an outbreak of flu-related purulent bronchitis and bronchopneumonia in the huge army camp at Etaples in France (winter 1916), and at Aldershot Camp in England (March 1917). In both instances, the morbidity was high and the mortality was around 25–50%, and it appears that during this period, the flu strain was slowly evolving from one of relatively low virulence to one of great virulence. There was then a brief lull at the end of the summer of 1918, but the pandemic began in earnest in late October.

A pandemic is defined as an epidemic that affects just about all people, and causes severe illness (and loss of life) during a period of one to two years. The influenza pandemic of 1918-1919 almost certainly rates as the greatest pandemic of all time. It affected people all around the World, from the Pacific (20% of the population of Western Samoa died) to Africa (5% of the population of Ghana died) to the USA (more than half a million Americans died). Two graphic accounts written at the time illustrate both the extent and rapidity of its effects. The first is taken from a letter written by a young physician at an army camp near Boston, Massachusetts:

> Camp Devens is near Boston, and has about 50,000 men, or did have before this epidemic broke loose… . This epidemic started about four weeks ago, and has developed so rapidly that the camp is demoralized and all ordinary work is held up till it has passed… . These men start with what appears to be an ordinary attack of La Grippe or Influenza, and when brought to the Hosp. they very rapidly develop the most vicious type of pneumonia that has ever been seen. Two hours after admission they have the mahogany spots over the cheek bones, and a few hours later you can see the cyanosis extending from their ears and spreading all over the face… . It is only a matter of a few hours then until death comes, and it is simply a struggle for air until they suffocate. It is horrible…We have been averaging 100 deaths per day.

This letter was printed in the British Medical Journal (December 1979), just a few months after the publication of a letter from C. Langton Hewer, who had served as a young man in the Royal Army Medical Corps in Europe at the time. He also recalled (British Medical Journal, January 1979):

> …the curious violet cyanosis which often occurred quite soon after the onset of the attack. This was quite different from the bluish-grey of ordinary cyanosis and also from the pinkish hue of carbon monoxide poisoning. The phenomenon was so remarkable and widespread that a medical journal sent an artist to make a watercolour sketch of one of our patients. The picture was duly reproduced but the patient died the same evening and the artist a few days later. The course of the disease could be extremely rapid. For example, it was quite common for two or three men to collapse during a morning route-march, to be carried back to barracks on stretchers, and to die before nightfall.

The high mortality associated with this pandemic was primarily due to the pneumonia, which so often followed the initial infection. This was sometimes caused by the influenza virus itself, but was more often due to the bacterium *Streptococcus pyogenes*. This invaded the already damaged lungs and produced a fulminating inflammatory condition, which led to respiratory collapse and death. With the advent of the sulfonamides in 1935, and subsequently, the penicillins and other antibiotics from 1945, the likelihood of such a high mortality rate from bacterial pneumonia in future pandemics was much reduced.

Historically, influenza has probably always afflicted mankind. Similar viruses affect domestic animals and various epidemics of 'catarrhal fever' were associated with epidemics of 'horse colds.' The close association between humans and horses in previous centuries may have allowed a vertical transmission of influenza from horse to man. However, there is convincing evidence that a similar transmission can occur from the pig (swine flu) to man, and the flu pandemic of 1918–1919 may have started in this way, although an avian origin is probably more likely. Evidence for the transmission of influenza from the pig to other mammals was first obtained by Richard Shope in 1932. He managed to infect various animals with swabs taken from nasal secretion of pigs suffering from swine flu. His proposition, that the pig could be a kind of 'safe haven' (reservoir) for the virus between human epidemics and pandemics, is widely accepted.

There is good documentary evidence of epidemics of the 'English sweat' sweeping through Britain in the years 1510, 1557, 1732, 1775, 1782, 1836 and 1847. The symptoms included a brief period of high temperature and sweating, followed by a discolouration of the skin and sudden death. This is highly reminiscent of the symptoms seen in the 1918 pandemic. Much better documentation exists from around the middle of the 19th century, and the pandemic of 1889–1892 appears to have been particularly severe. This also caused a high rate of mortality, especially amongst the 20–40 year age group, a feature that it shared with the 1918–1919 pandemic.

So what do we know of this disease, that the British physician Theophilus Thompson (a chronicler of influenza outbreaks from 1510 to 1852) described in 1852 as being 'of all epidemics the most extensively diffused and apparently the least liable to essential modification either by appreciable atmospheric change or by hygienic conditions under the control of man'? There are two main types of influenza virus – influenza A and B. Both are spherical viruses with a diameter of around one hundred nanometres (100 nM). The A-type virus is responsible for the world-wide pandemics in man, animals and birds. The B-type virus causes local epidemics, especially amongst school children. These major types are further classified into sub-types according to the two classes of glycoproteins on their cell surface: *haemagglutinin (HA)* and *neuraminidase (NA)*. At present, 15 discrete subtypes of HA and nine forms of NA have been identified, and these determine the 'antigenic signature' of the viruses, that is, the foreign proteins that are recognised by the host's immune surveillance system. The two proteins have key roles in the life cycle of the virus. Haemagglutinin assists in the docking of the virus particle with the host cell, which involves interactions between sugar residues of the haemagglutinin and so-called gangliosides (which are polysaccharides) on the surface of the host cell. One particular sugar residue of the ganglioside – neuraminic acid – is especially important in these interactions. Once bound through these multiple interactions, the host cell membrane encapsulates the virus and it is then internalised. A subsequent fall in pH results in a change in the three-dimensional structure of the haemagglutinin, which ultimately leads to rupture of the virus and release of viral RNA and enzymes. The neuraminidase has an equally important role, but for the release process rather than for uptake. It catalyses the cleavage of a neuraminic acid residue from the host cell ganglioside, and this results in a change in shape of the cell surface with the resultant release of new virus particles.

The influenza strain of 1918 appears to have had a haemagglutinin protein that was especially well adapted for human-to-human transmission and with a neuraminidase that allowed a specially efficient release of new virus particles. Certainly, the mortality was high and there was a unique age distribution. Usually, influenza produces a U-shaped mortality curve with excess death amongst the very young and the very old. The 1918 pandemic produced a W-shaped curve with a very significant additional peak in the 15–45 age range.

Clearly, a knowledge of the antigenic status of the virus and an understanding of the roles of these two glycoproteins are essential for effective drug design. Unfortunately, as with many viruses, mutation can occur and this results in changes to the cell surface glycoproteins haemagglutinin and neuraminidase. The antigenicity of the virus thus alters, and this is why someone

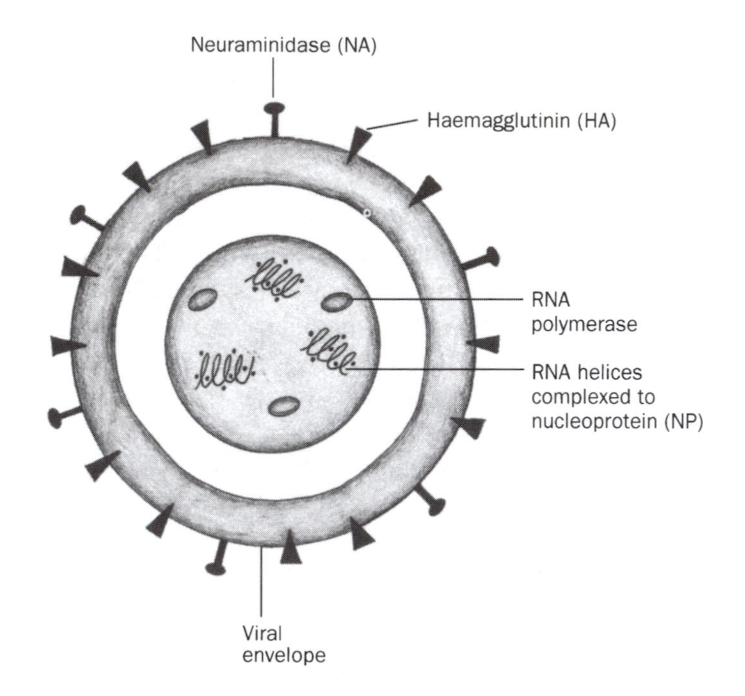

A flu virus

who has had a bout of flu in their youth may suffer again some years later. The World Health Organisation classifies the viruses according to their antigen status. Thus, the so-called Asian flu pandemic of 1957, which probably emanated from China, but was first recognised as an epidemic in Singapore in April 1957, was caused by a virus serotype called A/Singapore/57/H2N2 or simply H2N2. The earlier epidemics of 1946/1947 to 1957 involved the H1N1 serotype, while the Hong Kong flu of 1968 was H3N2. It is immediately apparent why it is possible to suffer from influenza on several occasions during a lifetime, and also why the young are particularly prone to contract the disease. The Russian flu of 1978 provides a good example of this phenomenon. This had the H1N1 serotype, and mainly affected those under twenty years of age, because this same serotype had been involved in the pandemic that ended in 1957.

The link with animal viruses was most dramatically demonstrated when several hundred military recruits at Fort Dix in Georgia, USA, contracted swine influenza in early 1976. The virus isolated from some of the recruits, termed A/New Jersey/H1N1, was found to be almost identical to Shope's swine flu virus, also A/H1N1. Since this was believed to be the causative agent of the great pandemic of 1918–1919, the US government of Gerald Ford panicked into initiating a major vaccination programme, not least because experts like Sabin and Salk (of polio virus fame) were in favour of the policy. On March 24th, 1976, Gerald Ford addressed the nation on TV and spoke of the very real possibility of "an epidemic of this dangerous disease next fall and winter", and went on to pledge the expenditure of $135 million "for the production of sufficient vaccine to inoculate every man, woman and child in the United States." The spectre of hundreds of thousands of dead Americans was too much for the US Congress, and they duly authorised spending for the campaign. This immediately precipitated a crisis within the pharmaceutical industry, because they feared, with some justification as it turned out, that production and administration of such vast quantities of vaccine would inevitably lead to numerous post-vaccination medical problems, and perhaps a number of deaths. They demanded that the US government indemnify them against possible insurance claims, and they eventually won the argument. There was much debate in Europe over the likelihood of another 1918-style pandemic, and the general view of the time is revealed in an editorial that appeared in the *Lancet* (July 3rd, 1976):

> The (Fort Dix) virus did not spread extensively – there were probably about 500 infections in the whole camp and no evidence of spread to civilians outside, even those in close contact with infected soldiers... . The virus was clearly capable of infecting man much more successfully than other antigenically similar viruses taken from pigs, but it seemed to be milder than a number of viruses isolated in the recent period of "drift", which have no pretensions to being possible pandemic strains.

Nonetheless, in August 1976, Ford signed a bill called the National Swine Flu Immunisation Program, and the great vaccination campaign began in the autumn of that year. As the pharmaceutical industry had predicted, there was a number of deaths, but more seriously, there was more than 1000 cases of Guillain-Barre syndrome, a condition in which there are serious neurological problems including total paralysis in some patients. Once the press realised the extent of these post-vaccination problems, the vaccination programme was doomed, and it came to a complete halt in December 1976. As many sceptics had predicted, the Fort Dix swine flu outbreak was an isolated incident, and not the forerunner of a great pandemic.

None-the-less, there is a recurrent fear that a new, more deadly strain of influenza will make the transition from an animal or bird host to man. The natural reservoir for the virus is the wild aquatic bird population, especially ducks, and one reason why so many of the new strains of influenza originate from the Far East, is that in this region, humans live in close proximity to aquatic birds and to swine. The virus appears to have adapted to the birds over a period of centuries. The genes for the various viral proteins are highly conserved and there are only small changes in the amino acid sequences of the associated proteins over many generations. This implies that there is an optimum association between the virus and its host – the virus is able to reproduce without causing significant illness or mortality within the aquatic bird population. However, when a strain of virus succeeds in making the transition into swine or perhaps directly into humans, there will usually be a rapid period of genetic adaption and perhaps enhanced virulence. There is a very limited supply of tissue samples from the 1918 pandemic – it is limited to two US soldiers and one Inuit woman whose remains were found in the permafrost of Alaska – and it had an H1N1 serotype. But interestingly, the amino acid sequence of the HA is much closer to that of typical avian HA than it is to HA from more recent swine or human flu strains. The X-ray structure of the haemaglutinin from one of these 1918 samples has been published recently (*Science,* January 2004) and revealed that the receptor-binding region, while avian in character, had been modified to allow it to bind more strongly to human receptors.

Humans can infect pigs, and pigs can certainly infect humans, and the pig may well be the animal of choice for the reassortment of viral genes between aquatic birds and humans – pigs certainly have cells with receptors for both avian and human influenza viruses. When a new strain is introduced into humans, it evolves rapidly and the greatest changes are observed in the gene that codes for HA. The influenza A haemagglutinin subtypes that are presently circulating are H1 and H3, and the recent flu seasons have involved versions of H1N2 and H1N3. However, the relative mildness of the infections in these seasons suggests that the dominance of the present strains may be on the wane, and the reappearance of an H2 strain could be serious since these more or less disappeared from humans in 1968 and no one under the age of 35 has any immunity.

A direct transition of influenza from birds to humans has occurred several times in recent years. In Hong Kong, a new strain of fowl influenza, H5N1, was transferred from chickens to humans during 1997. Thousands of chickens died from the infection, and between June and December, 18 people were infected and six of these died. This precipitated the slaughter of around 1.6 million chickens, but fortunately, there was no evidence that human-to-human spread of the new strain had occurred. This is a necessary prerequisite for an epidemic or pandemic to occur. More recently, in March 2003, there was an outbreak of the highly pathogenic H7N7 strain of avian influenza in poultry in Holland. Eighty slaughterhouse staff contracted viral conjunctivitis and one veterinarian died from a respiratory infection. Finally, in January 2004, there was an outbreak of H5N1 avian influenza in chickens in Vietnam, which soon spread to other parts of South-East Asia with more than 20 people infected.

Wild bird markets are probably the most important source of new influenza variants, and the intermingling of domestic chickens with aquatic birds has certainly resulted in the transmission of H5N1, H5N2 and H7N7 strains, all of which can be lethal for poultry and could make the leap into pigs and then humans. The recent emergence of the SARS virus was probably the result of just such a transmission of virus from a market animal – most likely a masked palm civet cat or a racoon dog.

SARS – or severe acute respiratory syndrome – first hit the headlines in March 2003, and was initially believed to be caused by a new strain of influenza. However, the coronavirus responsible probably made the transition from its animal host some time in 2002. Cases of an 'atypical pneumonia' had been causing deaths in the Guangdong Province of China since November 2002, but there had been little comment from the Chinese authorities. In February, outbreaks of what came to be termed SARS were reported in Hanoi and in Hong Kong, and this precipitated frenzied and ultimately successful control measures by the WHO. The symptoms were similar to those caused by influenza – fever, chills, headache and muscle pain – but after 2–7 days, patients developed a persistent dry cough and perhaps difficulty in breathing as the immune response to the virus damaged the lungs.

At the peak of the outbreak in early May 2003, 200 new cases a day were being reported and the mortality was as high as 50% in those over 65, although the overall rate for all ages was closer to 10%, and only 1% of those under 24 died. By July 2003, when the outbreak was declared over, 8439 cases had been reported and there had been 812 reported deaths. The situation was saved on this occasion because the virus required close interpersonal contact for its transmission, and was most typically spread by inhalation of droplets of virus-laden mucus. However, 321 victims in the Amoy Gardens apartment block in Hong Kong were apparently infected by virus in a faulty sewage system.

The coronaviruses are typically associated with the common cold, and are so-called because of their crown-like halo of protein spikes. Since they are RNA-(+)-viruses, they are able to code for their own m-RNA, which codes for the proteins involved in replication and can thus reproduce without much help from the host cell. On this occasion, the human race has been lucky and until the animal vector for this virus is identified, there will be the ever-present fear that the next time it jumps species, it may have an improved ability to both infect and subsequently kill its victims.

Influenza: the treatments

There are, of course, of number a vaccines available, which are used prophylactically for those at most risk: the elderly and those whose immune systems are compromised (patients receiving cancer chemotherapy, organ transplants, *etc.*). These are only useful against the known serotypes. The development of effective drugs is fraught with difficulties, not least due to the rapidity with which the infection strikes the patient. It is not uncommon for a person to be able to time, almost to the minute, when they first perceived the initial symptoms of malaise, chills and shivering, headache, muscle aches and perhaps a sore throat. Most patients feel the need to go to bed immediately, where they suffer from shivering, headache and backache. Sometimes there is an associated cough, but the whole period of prostration rarely lasts for more than three days. Recovery is usually rapid, although there may be associated respiratory infections, with pneumonia being the most serious sequel.

It is apparent that drugs must either be given prophylactically or immediately when the first symptoms are noted. Historically, two drugs have pre-eminence and these are amantadine and rimantadine. Both of these have very simple cage-like structures and are derivatives of adamantane. After the virus has attached to the cell via haemagglutinin and has been engulfed by pinocytosis, a change of pH leads to an initial uncoating followed by removal of the inner matrix proteins. These two drugs appear to interrupt the process whereby the structural protein on the inside of the viral envelope (matrix protein) is cleaved to allow the viral RNA and key enzymes to pass into the host cell cytoplasm. Amantadine has been in use since the mid-1960s (especially in Russia), and both drugs are active at the microgram level. A typical daily dose is around 200 mg, although rimantadine is about four times more potent; hence, it can be used in smaller amounts. There is the possibility of mild side-effects involving dizziness, loss of concentration and memory, but on the whole, the drugs are relatively well-tolerated and about 70–90% effective when taken prophylactically.

The influenza virus is an RNA-(−)-virus and possesses its own RNA polymerase enzyme. The complete RNA genome codes for eight proteins – two structural (matrix) proteins M_1 and M_2, haemagglutinin, and neuraminidase and four proteins involved in replication, three of which make up the polymerase enzyme. Substrates for this are the usual ribonucleosides and there has been some success with the use of nucleoside analogues as inhibitors of the enzyme. The drug ribavirin has been the most successful, although this has to be administered by aerosol. These days, its major use is for the treatment of infections caused by respiratory syncytial viruses (especially in children), since these can cause long-term morbidity.

But the brightest hope for influenza therapy is an Australian discovery, a family of drugs known as *neuraminidase inhibitors*. As the name suggests, these inhibit the actions of the viral neuraminidase, and since this is intimately involved with the budding of new virus particles, there is a consequent reduction in the infectivity of the virus. The three-dimensional structure of the enzyme was elucidated in 1993 using X-ray crystallography. This allowed the investigators to study how the neuraminic acid residue of the cell surface ganglioside fitted into the active site of the enzyme, and thus, how they might construct molecules to occupy the active site in its place. The Australian research group, led by Mark von Itzstein and Michael Pegg, at Monash University in Melbourne, went on to prepare several structural analogues of neuraminic acid that were very effective neuraminidase inhibitors. One of these (called zanamivir, Relenza) was prepared by Glaxo-Wellcome in the UK, and was approved for use in 2000. It is administered as a dry powder by inhalation, but was almost immediatedly eclipsed by another neuraminidase inhibitor (oseltamivir, Tamiflu) from Hofmann-LaRoche, which is effective by the oral route. Both drugs reduce the duration of the illness by 1.5 to 3 days if given within two days of the first symptoms of influenza; thus, there is a good chance that this Antipodean discovery may help prevent the next flu pandemic.

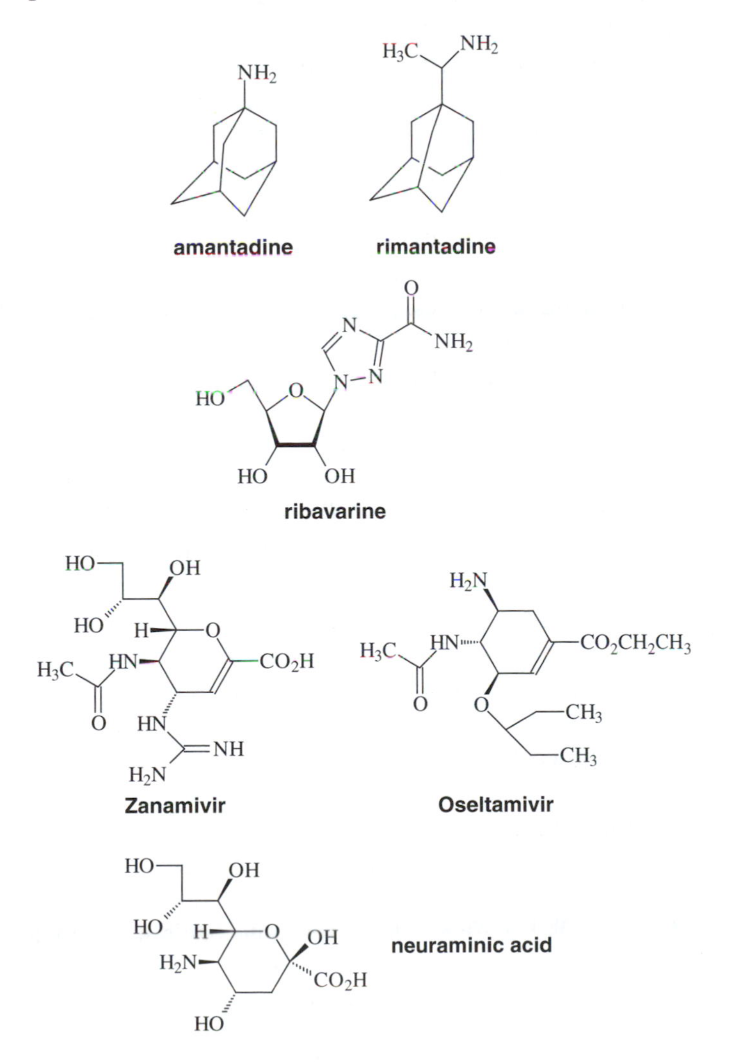

The Common Cold

About two-thirds of colds are caused by rhinoviruses – members of the picornavirus group (as are the polio viruses), and all of them are RNA-(+)-viruses. These have an icosahedral (20-sided, near-spherical) shape and their protein coat is made from four different proteins with widely differing amino acid compositions. It is not therefore surprising that there are more than 90 serotypes; hence, there is little possibility of a useful vaccine, and none of us has much chance of becoming immune to all forms of the common cold. Other viruses that produce the symptoms of a cold include coronaviruses, adenoviruses, coxsackie viruses, orthomyxoviruses, paramyxoviruses, respiratory syncitial viruses, echoviruses and enteroviruses.

Rhinoviruses find the human nose a near-perfect environment in which to grow. It has a temperature of 32–33°C (optimum for growth of the viruses), which is a few degrees cooler than the body temperature (37°), and an abundant supply of oxygen for rapid growth. The viruses do not thrive in the lower respiratory tract (lungs and bronchioles) because the temperature is higher, although bacteria like *Haemophilus influenzae* and *Streptococcus pyogenes* do thrive in this environment. The initial infection in the nose causes the classic symptoms of dry throat and runny nose, and is usually followed by a spread of infection to the pharynx, causing acute sore throat (pharyngitis), and to the larynx, causing hoarseness and loss of voice (laryngitis). Occasionally, the middle ear may also be involved and infection produces redness and perhaps earache. These are all virus-induced effects, and there is no possible utility for antibiotics, unless a secondary bacterial infection occurs.

The actual process of infection is not as simple as is usually believed. Most of the serotypes share a common receptor on host cells, which is an intercellular adhesion protein called ICAM-1. This is a transmembrane glycoprotein whose extracellular portion comprises several immunoglobulin domains of the IgG class. It appears that the normal role of ICAM-1 is to allow adhesion between the cell and leucocytes during injury or infection. Unwittingly, the ICAM-1 receptor has been hijacked by rhinoviruses as their receptor. Interestingly, ICAM-1 has other functions, which include sequestration of erythrocytes that have been infected by the malaria parasite *Plasmodium falciparum,* thus allowing them to avoid the attentions of the host's immune system.

There is little actual rhinovirus in saliva; hence kissing is not a normal route of transmission. The fine spray of virus particles produced in a sneeze is much more effective, especially if this is delivered to the interior of the nose. If the droplets have been deposited onto one's hand, then inserting a contaminated finger into the nose, or wiping an eye with the same finger, can cause spread of infection. Typical intermediary surfaces are doorknobs and loose change. An interesting experiment was conducted a few years ago in the USA, involving two teams of four card-players. One member of each team was infected with a rhinovirus, and while one team played normally, the other team wiped their cards with citric acid solution (a known virucide) before passing them on to the other members of the team. The team members who played normally all contracted colds, whilst the non-infected members of the other team remained cold-free.

As with influenza, the brevity of the incubation period and the fact that by the time the first symptoms appear, the virus will already be present in large amounts, means that prophylaxis and treatment are difficult. Add to this the fact that the cold may be due to another type of virus, for example, an adenovirus (a DNA virus) or a respiratory syncitial virus, and the problems appear insuperable. And indeed, at present, no drugs have reached the marketplace. The rhinoviruses are RNA-(+)-viruses and have two principal viral enzymes: an RNA polymerase and a protease, but although a number of potential enzyme inhibitory drugs have been evaluated in clinical trials, none has reached the marketplace.

Hence, the treatment for the common cold remains largely palliative rather than curative. Administration of anti-inflammatory agents like aspirin, decongestants like ephedrine, and anti-histamines will relieve many of the symptoms without necessarily reducing the period of infection. The codeine-like drug dextromorphan is also used and this acts upon the *N*-methyl-D-aspartate (NMDA) receptors in the brain to increase the threshold for cough induction. One new experimental treatment involves the intranasal administration of ICAM-1 fragments to saturate the receptors.

The use of megadoses of vitamin C was championed by the late Linus Pauling in his books *Vitamin C and the Common Cold* (1970) and *Vitamin C the Common Cold and the Flu* (1976). He advises:

> At the first sign that a cold is developing…begin the treatment by swallowing one or two 1000 mg tablets. Continue the treatment for several hours by taking an additional tablet or two tablets every hour.

Unfortunately, although a considerable amount of research has been carried out, there is very little support for the efficacy of such a regime. In addition, vitamin C is not cheap (when purchased through a pharmacy), and may also crystallise in the kidneys possibly causing problems. There is a similar lack of evidence for the efficacy of various preparations containing zinc, though zinc ions do inhibit the growth of rhinoviruses when these are grown in culture. In the absence of a 'magic bullet', perhaps the best, and certainly cheapest treatment, is to blow a gentle stream of warm air into the nostrils, since this raises the temperature of the nose above the optimum for growth of the rhinoviruses!

HERPES: COLDS SORES AND CHICKEN POX

If rhinoviruses and influenza viruses are the most common agents of infection in the developed world, then the herpes viruses run them a very close second. Very few people escape the chicken pox of infancy or the cold sores of adulthood. The former is caused by varicella zoster virus (VZV) and the latter by herpes simplex-1 virus (HSV-1). An increasingly prevalent virus – herpes simplex-2 virus (HSV-2) – is the predominant cause of genital herpes. These are all members of a family of viruses that also includes the potentially more pathogenic cytomegalovirus and the Epstein-Barr virus. The herpes viruses are much more complex than the rhinoviruses, and contain a relatively large, double-stranded DNA molecule within their icosahedral capsid, which is itself enclosed within a lipoprotein envelope. The DNA carries the blueprint for around 100 small proteins, three of which have major importance. These are a DNA polymerase, a protease and (for HSV-1, HSV-2 and VZV) a thymidine kinase. This last enzyme is of particular significance for the design of selective drugs, because other viruses and host cells possess kinases with very different properties.

The symptoms of infection by varicella zoster are usually unmistakable: the slight fever accompanied by a rash, which tends to be concentrated on the central part of the body (the trunk) rather than on the face, arms and legs. The spots (vesicles) resemble drops of water on the skin, and these soon dry to form scabs. The infected child is inconvenienced rather than incapacitated by the infection. In adults, chicken pox can have very serious consequences, and may lead to potentially life-threatening encephalitis or pneumonia. However, the more common form of varicella zoster infection in adults leads to a condition known as shingles. This is manifested as severe inflammation of the sensory nerves, which is usually accompanied by intense pain, and later by a rash. In most instances, it seems that the varicella zoster virus of infancy has lain dormant for many decades in the nerve cells, and has been reactivated in response to some kind of trauma (perhaps depresssion or an infection), to produce an attack of shingles.

A similar kind of dormancy between attacks is seen with HSV-1 and HSV-2 infections. Most people first encounter HSV-1 in childhood when kissed (for example) by an infected adult. In crowded conurbations, as in a town or city, well over 90% of the population may carry the virus. The virus generally causes the familiar cold sore, which begins with a small area of redness (and a tingling sensation) on the lip, and rapidly evolves into a cluster of tiny vesicles, and thence into a pustule that eventually scabs. The liquid exudate from the pustule contains highly infective virus particles. A cold sore may also occur inside the nostril or within the outer ear, and usually recurs in the same place over many decades. In between infections, the virus appears to lie dormant in nerve cells at the top of the spine.

HSV-2 causes much more discomfort since it usually infects the genital surfaces, especially the labia minor and clitoris in women, and the tip of the penis in men. In addition, the anus may be affected in both sexes, and there is strong evidence that the prevalence of genital herpes has increased dramatically during the past 30 years as the population has become more promiscuous. It is probable that an increased awareness of the dangers of spreading HSV-1 to infants has led to a decrease in HSV-1 infection. This in turn resulted in a greater number of young people with no antibodies to HSV-1, which at least provided some protection against HSV-2. By the late 1980s, more than 50% of young adult males in the larger cities of the USA (and probably a similar percentage in Europe) were infected with HSV-2, and this figure would have risen still further if the AIDS crisis had not produced a greater awareness of safe sexual practices.

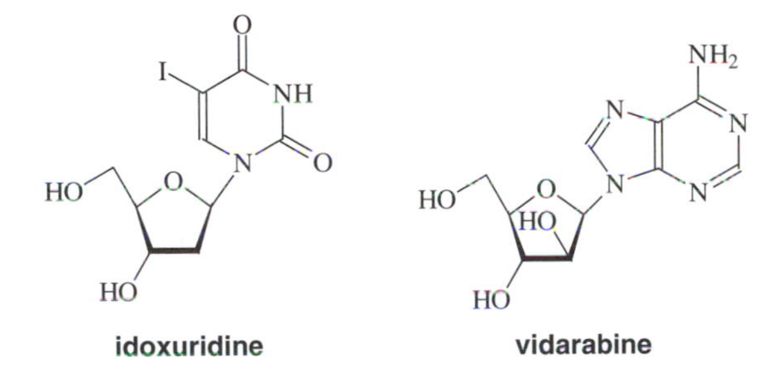

The treatment of most herpes infections involves drugs that inhibit one or other of the viral enzymes. Initially, a number of nucleoside analogues (idoxuridine, vidarabine) were developed for topical use – they are all inhibitors of viral DNA polymerase, but are too toxic for internal use. However, the major breakthrough occurred in the laboratories of the pharmaceutical company Wellcome in North Carolina in 1977. As will be described in the next chapter, the research group of Gertrude Elion and George Hitchings had been working since the mid-1940s on inhibitors of the various enzymes of DNA metabolism, and had already discovered a number of anti-cancer agents – they subsequently shared the Nobel prize for Physiology or Medicine in 1988. Their new drug, itself a modified nucleoside subsequently called *acyclovir* (Zovirax), had not only a very high potency as an antiviral agent but also great selectivity. (The antiviral activity was actually first demonstrated at the Wellcome Labs. in Beckenham in 1974, but the synthetic work was carried out in the USA.) Acyclovir inhibited the growth of HSV-1 and HSV-2 as well as VSV, and also had some activity against Espstein-Barr virus; but intriguingly, it had no activity against other DNA and RNA viruses and no toxicity towards host cells. Clearly, the compound had all of the attributes of a 'magic bullet'.

Its mode of action was subsequently elucidated through the use of the drug labelled with the radioactive isotopes (^{14}C and tritium). In infected cells, three new radioactive products were obtained: the monophosphate, diphosphate and triphosphate of acyclovir. The sequence of events (see Fig. 3.1) was shown to involve the thymidine kinase enzymes of the herpes viruses. These catalyse the attachment of the first phosphate group, and then two host cell kinases assist with the attachment of the second and third molecules of phosphate. The final product, acyclovir triphosphate, was an extremely potent inhibitor of viral DNA polymerase – it prevented the assembly of the chain of nucleotides that formed the backbone of viral DNA. The crucial finding was that acyclovir was a substrate for viral thymidine kinase but could not interact effectively with host cell thymidine kinase, hence, only the virus could produce the actual inhibitor acyclovir triphosphate. The host cell

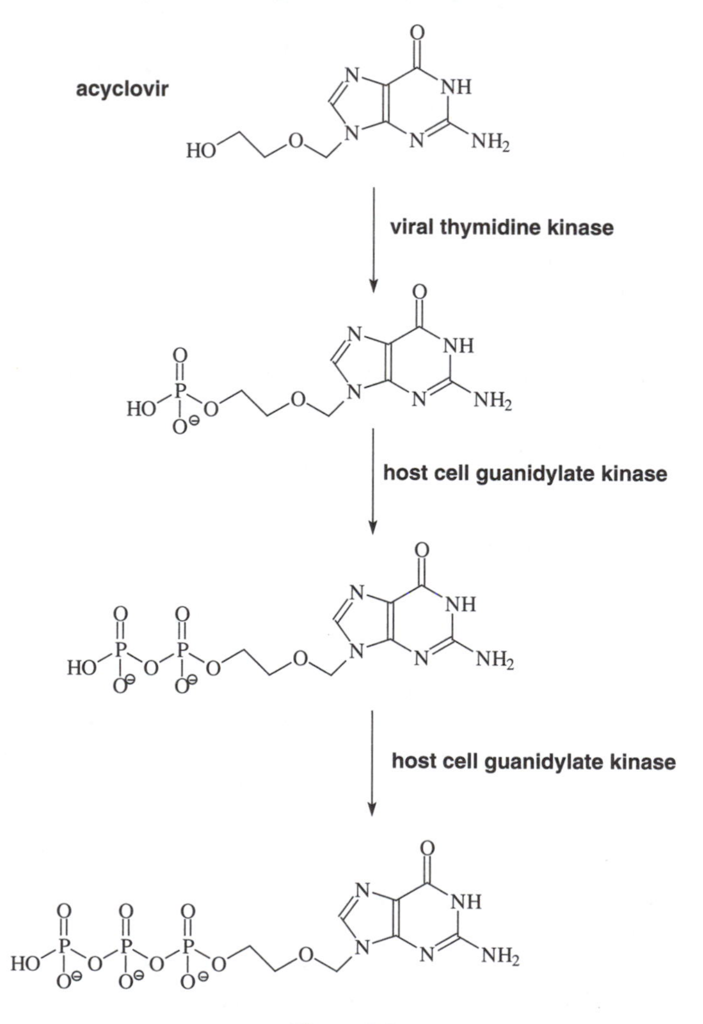

Figure 3.1

DNA polymerases of uninfected cells were thus unaffected, even if ACV entered the cells.

Acyclovir rapidly became Wellcome's leading drug in terms of worldwide sales. It was (and still is) used topically as a treatment for cold sores, and internally (by the oral route or by intravenous administration) for the treatment of recurrent genital herpes, and for shingles. It has been particularly valuable for the treatment of immune-compromised patients (receiving cancer chemotherapy or immunosuppressive drugs following an organ transplant), keeping them free of potentially lethal herpes infections.

A range of analogous drugs have appeared since 1977, and these include ganciclovir (from Syntex, 1980), which has better anti-viral activity against cytomegalos virus, but is too toxic for general use; penciclovir (Beecham, 1993); and famciclovir (also Beecham, 1994), which is a prodrug of penciclovir. Some resistance to these drugs, especially to acyclovir, has been observed, and this appears to be due to reduced levels of thymidine kinase in

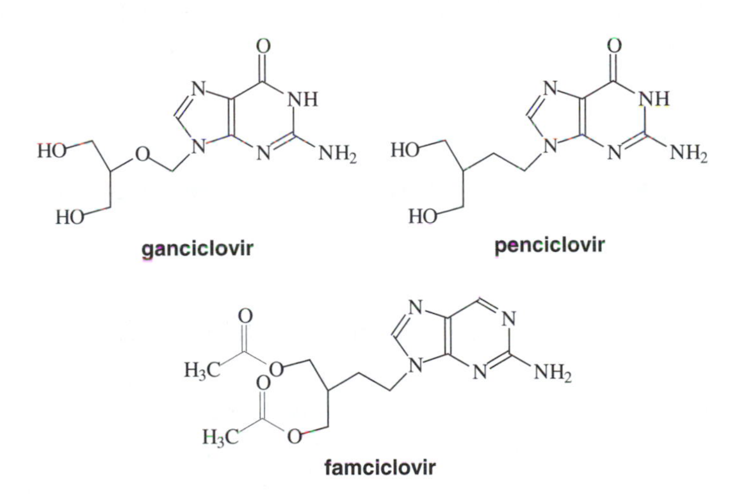

the virus, or due to alterations in the structures and activities of the viral thymidine kinase or DNA polymerase.

There is still no really effective drug for the treatment of cytomegalovirus or Epstein-Barr virus. Most people become infected with CMV at some stage in their lives without experiencing any symptoms, but in immunocompromised patients, the infections are potentially life-threatening. Indeed, in AIDS patients, CMV infection is one of the most common causes of death. The Epstein-Barr virus is most commonly encountered as the causative agent of infectious mononucleosis or glandular fever. This is most prevalent in adolescents, and is passed on by kissing and other forms of sexual activity. Although this is usually a relatively mild condition, there is good evidence to suggest that the virus is retained and may be a contributory factor in the development of certain forms of cancer. In Africa, where most people have immune systems weakened by malaria, the Epstein-Barr virus is certainly the causative agent of Burkitt's lymphoma, a cancer of the lymph glands in the neck. In the non-malarial world, it may contribute to the development of Hodgkin's disease (another cancer of the lymph glands) and nasopharyngeal carcinoma. Clearly, effective drugs for the treatment of these two viruses are urgently needed.

Notwithstanding these deficiencies, there is no question that acyclovir has had a major impact on the treatment of herpes infections. It is now on sale, in most countries as an over-the-counter (non-prescription) drug and as Gertrude Elion stated in her address to the Nobel Committee in 1989:

> In addition to the clinical utility of ACV, the lessons learned from its development have proven to be extremely valuable for subsequent antiviral research. In depth studies of mechanisms of action have led to a better understanding of the enzymatic differences between normal and virus-infected cells and have given impetus to the search for other viral-specific enzymes which are capable of therapeutic application.

It is interesting to note that the financial fortunes of Wellcome were completely dominated by this drug for many years, and its supremacy was eventually supplanted by another Wellcome drug – azidothymidine or AZT – the first drug to show efficacy against the most infamous virus of the late 20-century – the human immunodeficiency virus or HIV.

AIDS: HISTORY OF THE DISEASE

It was perhaps inevitable that the 'sexual revolution' of the 1960s would increase the incidence of sexually transmitted diseases. What no one could have predicted was the appearance of a disease as deadly as AIDS. The first evidence that a new disease had appeared came from the University of California Medical Center in Los Angeles. During the latter part of 1980 and the early part of 1981, five young men died as a result of the combined effects of pneumonia caused by the parasitic protozoan *Pneumocystis carinii*, and generalised infections caused by the yeast *Candida albicans* (which causes thrush) and cytomegalovirus. Normal healthy individuals are exposed to these organisms throughout their lives, but are usually kept free of disease by an active immune system. These five men had virtually no capacity to raise an immune response, because they were almost completely devoid of T-lymphocytes, and this was a situation that had only previously been seen in patients undergoing extensive chemotherapy for cancer. There were two other common features: all five men were homosexual and had been highly promiscuous, and had taken amyl nitrite 'poppers' as sexual stimulants. Later in 1981, reports began to appear from New York, Los Angeles and San Francisco of young gay men dying from Kaposi's sarcoma, a very rare skin cancer that had hitherto only been seen in patients over the age of 50 years. These men were also deficient in T-lymphocytes.

By September 1981, the Center for Disease Control (CDC) in Atlanta had compiled a list of just over 100 deaths ascribed to Kaposi's sarcoma or to *Pneumocystis carinii* pneumonia or both. Ninety-five of these patients were gay white males. The number of such deaths rose to 270 by the end of the year, and all of the post-mortems revealed immune systems that were essentially devoid of T-lymphocytes with body organs engulfed by organisms like *Staphylococcus aureus, E.coli* and *Cryptosporidium.* Somewhat surprisingly, the numbers of B-lymphocytes, that is, the ones already committed to producing antibodies, were within normal levels, and so the disease was clearly highly specific for the T cell system. In particular, one class of lymphocyte, the CD4 lymphocyte, was absent or very severely diminished in numbers.

A few isolated cases of the disease began to appear in Europe, with around half of these reported in France, and in 1982, the French Government set up a task force to trace the cause of the disease. In New York City, a growing number of non-gay, drug abusers were diagnosed with the condition, and in mid-1982, there was an outbreak amongst Haitians (including women) newly arrived in the city. These too were heterosexual, and had no history of drug abuse. There were, nonetheless, strong suspicions that the disease was sexually transmitted, and it was given the name Gay-Related Immunodeficiency Disease or GRID. Despite the obvious seriousness of the situation, the Reagan administration's animosity towards the gay community led to a 'gagging order' being imposed on the then Surgeon General, C. Everett Koop, and hence the American public was initially slow to learn about GRID.

This ignorance led to alarm as the first cases appeared in haemophiliacs, who had received the blood protein factor VIII. This was not only a serious health concern but also a major commercial problem, since worldwide sales of blood products was estimated to be around two billion US dollars. For haemophiliacs, the situation was dire. Because of the large quantity of blood plasma that had to be processed in order to obtain factor VIII, and the large number of injections they needed each year, it was estimated that each patient could be exposed to the blood of up to three million donors. At this time, the manufacturers of factor VIII did not use heat-treatment procedures, as they did from 1987, but nonetheless, the causative agent appeared to be pretty rugged. The appearance of the disease in haemophiliacs rendered the name GRID inappropriate for the condition, and this was now replaced by the term Acquired Immune Deficiency Syndrome or AIDS.

There was still no real awareness of the potential seriousness of the 'epidemic.' By March 1983, a mere 1350 cases of AIDS had been reported to the

Center for Disease Control (CDC) in Atlanta, and all but a few of these were in the USA or Haiti. The Republican-controlled US Senate and the White House of Ronald Reagan were reluctant to recognise a condition that affected gays and Haitians! The small number of European cases, most of whom had spent some time in Africa, seemed to imply an African origin for the infectious agent, with onwards transmission via Haiti to the USA.

One of the people who was to play a key role in the AIDS story, Robert Gallo of the National Cancer Institute in Washington, first became aware in late 1982 of this new infectious agent that devastated the immune system. He was immediately impressed by the similarity of effects to those induced by the human T-cell leukaemia virus, HTLV-1, which he had recently discovered, and which also ravaged the T-cell system. On the other side of the Atlantic, at the Pasteur Institute in Paris, a group led by the virologist Luc Montagnier were also wondering about the similarities with HTLV. In January 1983, they received lymph node tissue from a French AIDS patient and were able to demonstrate biological activity associated with the viral enzyme reverse transcriptase. At that time, only two retroviruses (that is, ones that contain RNA rather than DNA and need the enzyme reverse transcriptase to catalyse the production of viral DNA from viral RNA) were known, and these were HTLV-I and HTLV-II. Since Robert Gallo was the expert in this field, it was natural for Luc Montagnier to communicate his results to Gallo. The latter soon demonstrated reverse transcriptase activity in US samples from AIDS patients, and became convinced that HTLV was the infectious organism. This was entirely reasonable since the virus was endemic in the Caribbean and was also common in Africa.

Throughout 1983, there was intense competition between these two groups, and to a lesser extent with the group of Jay Levy in San Francisco, to be the first to isolate and identify the causative agents of AIDS. The sense of urgency was engendered by the fact that an increasing number of cases of heterosexually transmitted AIDS and mother-to-baby transmissions were reported during the year. There was also an alarming rise in the number of cases associated with needle sharing. The general public became alarmed about the possibility of infection from toilet seats, *etc.*, and even informed opinion began to talk of a 'plague' of AIDS. In an editorial in *Nature* (April 28th, 1983) with the headline 'No need for panic about AIDS', the editor John Maddox wrote:

> There is now a serious danger that alarm about the disease physicians call acquired immune deficiency syndrome (unhelpfully, AIDS for short) will get out of hand. For the characteristics of this previously unrecognized and perhaps non-existent condition are so alarming that the temptation to portray it as a disease invited by a decadent civilization – a kind of latter-day version of the fate of Sodom and Gomorrah – is almost irresistible.

His contention that "the notion that immune deficiency is acquired by means of a transmissible agent, a virus for example, is open to dispute, or at least unproven" was soon overturned by the rapidly emerging results from Paris, Washington and also from the group of Max Essex at the Harvard University School of Public Health. In the May 20th issue of *Science,* there were no fewer than five reports from these groups. The results, while appearing to confirm that an HTLV-like virus was involved, were confusing in the sense that the NCI group had isolated HTLV-I from an AIDS patient, while the virus isolated by the French group was neither HTLV-In or HTLV-II. Gallo also noted that, while HTLV-I induced T-cell leukaemia, the latency period was years rather than a number of months – the apparent latency period for development of AIDS. Even more disconcerting was the lack of AIDS in Southern Japan, where around 25% of the population had antibodies to HTLV-I, and hence had encountered the virus.

In September, the French group announced the isolation of a new virus, which they called lymphadenopathy-associatiated virus or LAV, from patients who were suffering from the early stages of AIDS. The virus appeared to have a particular affinity for T-cells that possessed CD4 receptors. LAV also appeared to be completely different from HTLV-I. In May 1984, the Gallo group reported the discovery of a retrovirus, from AIDS patients, which they called HTLV-III. A British group under Robin Weiss at the Chester Beatty Labs. in London confirmed, in December 1984, that HTLV-III and LAV were apparently the same organism. Ultimately, this came to be known as human immunodeficiency virus (HIV-I). Meanwhile, Levy had isolated a slightly different virus from his patients in San Francisco, and this he christened ARV (AIDS-associated retrovirus). This also proved (from analysis of its RNA and DNA composition) to be very similar to HIV-I, and was ultimately shown to be closer to the original wild-type strain rather than the forms isolated by Montagnier and Gallo Their viruses had been maintained in culture for so long that they had apparently mutated into a slightly different form of the virus.

Just as these momentous discoveries were being reported in the medical journals, the African epidemic of AIDS began in earnest. This disease was spread primarily via the heterosexual route, and through the multiple use of syringe needles for administration of medication. Since the advent of relatively cheap antibiotics and other drugs, the favoured route of administration in African countries had been by syringe. The disease spread from Uganda, Tanzania and Zaire to the whole of Southern Africa, and was usually rapidly fatal. Many of the early European cases of the disease were in people who were of African origin, and it began to appear that AIDS might be endemic to parts of Africa. The probable African origin of the disease was further supported by the discovery of an HIV-like simian virus (SIV) in rhesus macaques, and a hypothesis evolved that SIV had evolved to produce HIV. This became even more plausible following the simultaneous discovery of HTLV-IV by Max Essex of Harvard and LAV-II by Montagnier in March 1986. These new viruses were shown to be identical and were eventually called HIV-2. What was particularly striking about this virus was that it was endemic in West Africa and was 75% identical in its composition to one form of a primate (simian) immunodeficincy virus – SIV.

As more and more data became available, the researchers discovered that simian immunodeficiency viruses (SIVs) were present in just about all species of African monkeys, and the homology with HIV-2 was always good. There was an especially close similarity between the genomes of SIV from the sooty mangabey (a native of West Africa) and HIV-2 (endemic in West Africa). These viruses also attacked cells which had CD4 receptors on their surface, and it seemed likely that these viruses could have infected humans as man destroyed the natural habitat of the primates and they were forced to forage in the vicinity of human settlements. Initially, the human versions of SIV were relatively non-virulent, but through mutation, more dangerous variants like HIV-2 and later HIV-1 had appeared. Comparisons of the genomes of HIV-1, HIV-2 and various SIVs suggested that these viruses could have diverged as recently as the mid-1940s.

Supercomputers were subsequently used to analyse the genome (RNA and DNA sequences) of the various strains of HIV-1, and this revealed that there were several main types. Type A was found in South Africa and India; type B in North America, South America and Thailand; whilst type D was the most dangerous, and this was found primarily in Rwanda, Uganda and Tanzania. The genetic variance within the eight main types was never more than 30%, and if one assumed a mutation rate of 1%, then the initial appearance of HIV-1 could be dated to the early 1960s, with a dramatic increase in virulence during the early 1970s. The prevalence of African wars during this period with the inevitable human migrations, rape and prostitution would have hastened the spread of the new virus, as would the availability of new injectable antibiotics and other drugs and the consequent multiple use of syringe needles. In America and to a lesser extent in Europe, the availability of cheap foreign travel coupled with the promiscuity and drug abuse of the 1970s, ensured the spread of the virus from Africa.

In the February 5th, 1998 issue of *Nature*, David Ho and co-workers reported an analysis of what purports to be the oldest sample of HIV. This was isolated from a tissue sample taken from a Bantu adult male who died in 1959, in the city of Leopoldville (now Kinshasa), which was then in the Belgian Congo. The DNA sequences of various gene fragments were shown to be very close to those of ancient HIV subtypes, especially to the D subtypes. The rate of evolution of HIV from its putative ancestral progenitor has been estimated, and it would appear that the early subtypes A to D evolved from a single HIV infection sometime in the 1940s.

Others were less convinced by the scientific hypotheses that were pre-eminent in the mid-1980s. The possible involvement of the CIA with a genetically engineered virus or a 'viral invasion' from outer space were both suggested. More seriously, the American virologist Peter Duesberg of the University of California at Berkeley claimed that HIV was not the causative factor of AIDS, but that it was due to a degenerate lifestyle – what he termed a 'self-destructive gay lifestyle'. His arguments were appealing, especially to the media, and the London *Sunday Times* championed his cause for several years. His main contention was that AIDS did not have the typical characteristics of an infectious disease, because "viruses and bacteria work fast or never." In addition, even with the sensitive assay methods that had been developed, it was usually impossible to demonstrate the presence of the virus in more than 1 in 400 of peripheral lymphocytes. It was difficult to perceive how such a low level of infection could predispose to AIDS. What he failed to take into account was the possibility that the main reservoir of infection might be cells in the bone marrow, lymph nodes, *etc.*

Duesberg drew special attention to the situation in the USA (up until 1990), where the disease had been confined almost exclusively to men (more than 90% at that time) who were gay, drug users or haemophiliacs. The long latency period (up to 10 years) was also incompatible, he believed, with a disease of viral origin. His preferred option was that a lifetime of drug abuse and/or gay promiscuity would predispose an individual to infection by many organisms as their immune systems became progressively immunocompromised. HIV was thus but one organism amongst many, *e.g.*, cytomegalovirus, hepatitis, HTLV, amoebae, candida, mycoplasma, *etc.*, leading to the condition of AIDS.

Many discoveries over the intervening years have provided a comprehensive refutation for his views. Probably the first of these was announced in the September 7th, 1995 issue of *Nature*. A group at the University of Oxford and the Oxford Haemophilia Centre followed the fates of 6278 male haemophiliacs diagnosed with the condition between 1977 and 1991. Of these, 1227 were unfortunately infected with HIV-1 between 1979 and 1986 through treatment with contaminated blood products. The researchers found that amongst 2448 men with severe haemophilia, the annual death rate was 8 per 1000 during the period 1977 to 1992 for those not infected with HIV, but rose steadily (and reached a peak of 81 per 1000 in 1991–1992) for those who had been infected. Amongst a group of 3830 men with less severe haemophilia, the corresponding figures were 4 per 1000, rising to 85 per 1000 (in 1991–1992), for those who had been infected. None of these men were reported to have been gay or to have used drugs, hence, this part of Duesberg's hypothesis was clearly untenable.

The apparent elusiveness of HIV in circulating lymphocytes had been explained two years earlier. In March 1993, two groups reported, quite independently, that during the long latency of AIDS, the virus can be found in the lymph nodes, spleen and tonsils of patients. Nor was it quiescent, because they provided evidence that it reproduces itself in these tissues. The apparent absence of HIV in circulating T-lymphocytes is then explained by the fact that once the virus leaves the lymph nodes, it seeks out and rapidly destroys any T-cell carrying the CD4 receptor on its surface. As many as one billion CD4 positive cells are newly infected each day and a further billion dead cells are removed from the circulation. As the group of Pantaleo and Fauci succinctly stated in their *Nature* paper of March 25th, 1993:

> "We have demonstrated that during clinical latency, HIV accumulates in the lymphoid organs and replicates actively despite a low viral burden... . The peripheral blood does not accurately reflect the actual state of HIV disease."

The problems that this presented for those who were desperately trying to devise therapies was clearly stated in the report by Embretson and co-workers in the same issue of *Nature*:

> "There are large numbers of latently infected $CD4^+$ lymphocytes and macrophages in which HIV can persist and be disseminated despite immune surveillance... . (and this) underscores the difficulties in designing treatment programmes... . Equally difficult is the problem of designing a vaccine to interdict the spread of HIV by a 'Trojan horse' mechanism, concealed inside covertly infected cells".

But a treatment had to be found. By 1991, the World Health Organisation estimated that HIV had infected around ten million people worldwide and they predicted a figure of 30–40 million by the year 2000. This has proved to be an underestimate, and figures released by the UN AIDS Survey in December 2003 suggest that 34–46 million people have HIV/AIDS of which 31–43 million are adults and 2–3 million are children. In 2003 alone, more than 5 million people were newly infected with HIV and around three million people died from AIDS. During the past 15 years or so, vast sums of money have been poured into research that will hopefully reverse the apparently inexorable progress of the disease, and there has been some real progress.

AIDS: PREVENTION AND TREATMENT

The lifecycle of HIV is shown in Fig 3.2. Initial interaction between the viral coat glycoprotein gp120 and the glycoprotein called CD4 on the T-lymphocyte or macrophage is followed by an additional binding to a co-receptor called

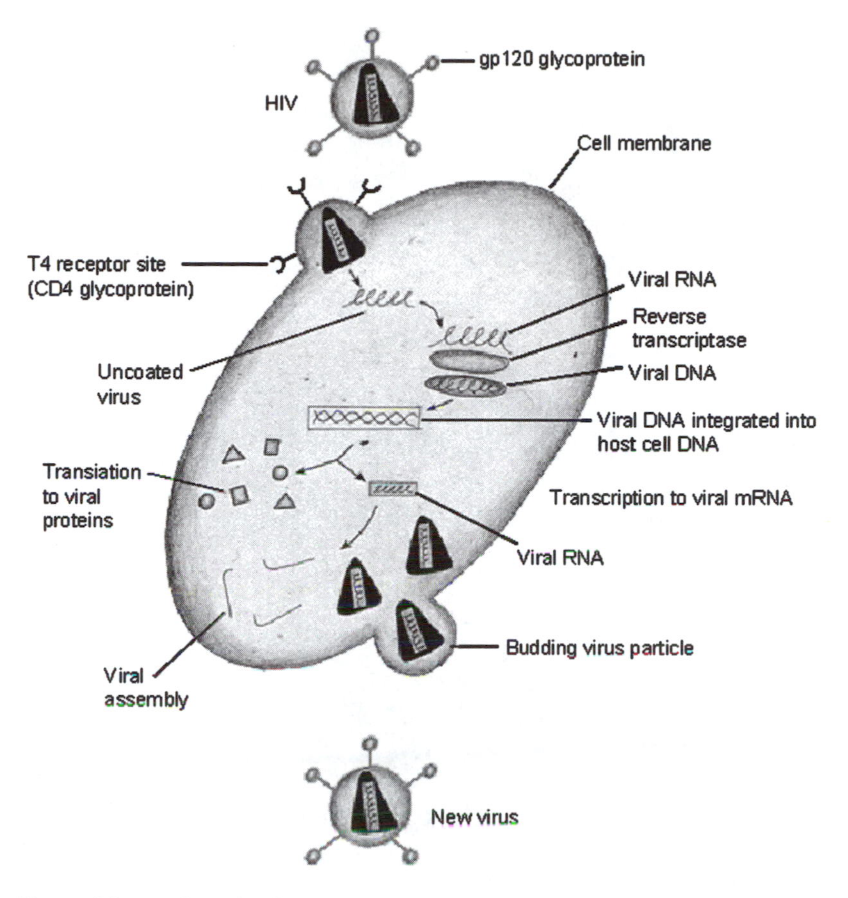

Figure 3.2a *Life cycle of HIV*

CCR5 (on macrophages) or CXCR4 on T-lymphocytes. These associations trigger a conformational change that brings another viral glycoprotein, gp41, 'crashing' down upon the host cell membrane, leading to fusion of the virus and host cell, and then penetration of the virus through the cell membrane. In the early stages of infection, it is infected monocytes or macrophages carrying the CCR5 receptor and the CD4-glycoprotein that predominate, while infected CD4-positive T-lymphocytes carrying the CXCR4 receptor are most prevalent during the long latency period prior to the development of AIDS.

Once inside the host cell, the virus loses its coat and the two single (+)-strands of viral RNA are released together with the enzymes viral reverse transcriptase (RT), integrase and protease. RT is an RNA-dependent DNA polymerase and is needed to catalyse the transcription of the genetic information present within the RNA into viral DNA, and thence production of double-stranded DNA. The new pro-viral DNA is now incorporated into host cell DNA using the enzyme viral integrase, and this incorporated DNA

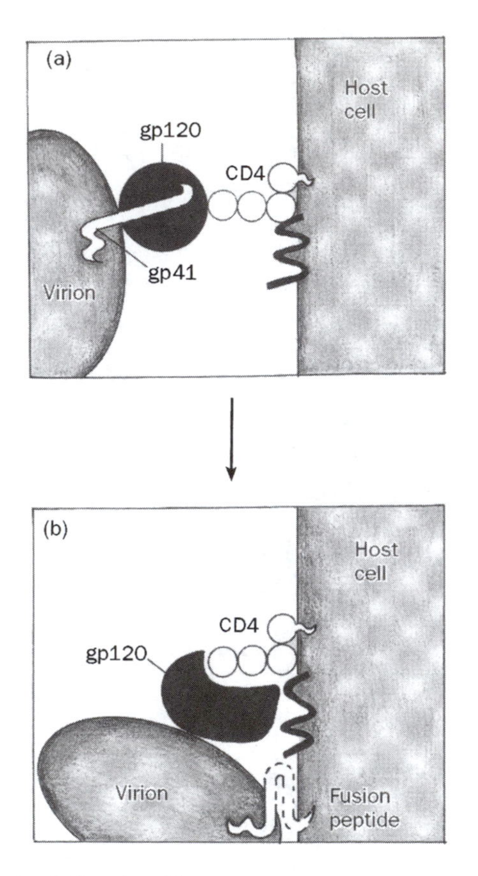

Figure 3.2b *Action of gp120 and gp41*

may now lie dormant for long periods. However, during the early stages of infection, the pro-viral DNA is immediately transcribed into viral mRNA by host cell RNA polymerase. The mRNA then codes for the production of new viral enzymes and coat glycoproteins. These are modified by viral protease before incorporation into the new virions along with the new copies of viral RNA. Finally, excess carbohydrate is trimmed from the newly created glycoprotein envelope using host cell glycosidases, and the mature virions are released from the host cell.

The immune system does mount a vigorous response in these early days. Macrophages ingest the viral particles and degrade them, and then display small portions of the virus (small peptides called epitopes) on their surface, and this attracts the attentions of helper T-cells, which in turn activate killer T-cells and B-lymphocytes. The killer T-cells attack cells displaying the viral epitopes, and the B-cells produce dedicated antibodies that also seek out and destroy such cells. Not surprisingly, in the early days after infection, around 30% of patients exhibit typical symptoms of their immune response to viral attack: fever, rash and swollen lymph nodes. Most patients then attain a relatively symptom-free phase during which the virus is sequestered in the lymph nodes, tonsils, *etc*. Despite the apparent latency of infection, a billion new virions are produced each day by newly infected CD4-positive cells and cleared through the circulation.

Over a period of time, these CD4-positive T-cells are destroyed, as much by the activated killer T-cells as by the virus itself, and the loss of these cells is evidenced by the lowering of the numbers of detectable CD4-positive cells. These should normally number around 800–1200 per cubic millimetre of blood, but this count drops initially to around 500 in HIV-positive patients and then to less than 200 in patients with full-blown AIDS. This process may take a few years, but typically takes around 10 years.

The variability in the rate of progression and also the near-impossibility of producing a multi-purpose vaccine derives from the fact that the virus continually changes the structure of its glycoprotein coat, thus making recognition and destruction by an ever-weakening immune system almost impossible. HIV is, in fact, one of the most variable viruses known, and this is believed to be due to the inefficiency of its enzyme reverse transcriptase. Every time this makes a copy of DNA, it makes at least 5–10 mistakes. Obviously, many of these changes will be either neutral or detrimental to the viability of the virus, but such is the rapidity of the rate of change, that statistically, new and viable strains of virus will be produced relatively frequently. For example, it has been estimated that the *env* gene of the virus is changing at the rate of about 1% per annum. This codes for the coat glycoprotein gp160 which is cleaved to produce gp120 and gp41; hence, these mutations inevitably lead to alteration in the constitution of these important glycoproteins. Interactions with T cells and other cells will obviously change.

Chemotherapy has always been the mainstay of AIDS treatment and this is likely to remain the case for some time to come. Historically, inhibitors of viral reverse transcriptase were the first to be used clinically, and AZT (azidothymidine, Retrovir, zidovudine) from Wellcome has become justly famous for providing the first glimmer of hope in the treatment of this seemingly untreatable disease. Like acyclovir, AZT requires phosphorylation by host cell kinases before it becomes an irreversible inhibitor of reverse transcriptase. AZT was first prepared in 1964 by J. P. Horwitz at the Michigan Cancer Foundation as a potential anti-metabolite for cancer chemotherapy, but was never used for this purpose It was first administered to AIDS patients in July 1985, and there was an immediate improvement in their immunological and clinical condition. A rise in their CD4-positive cells was also apparent. The results were so encouraging that Wellcome started a multi-centre trial in February 1986 and the drug was approved for prescription use in March 1987. The side-effects were serious: the nausea, vomiting, liver dysfunction and most bizarrely – bluish pigmentation of the nails were tolerable, but the myelosuppression (loss of white blood cells) often necessitated cessation of drug therapy. Resistance to AZT was also soon observed. Other similar drugs have been introduced more recently: ddC (dideoxycytidine, zalcitabine) in 1988; ddI (dideoxyinosine, didanosine); d4T (didehydrothymidine, stavudine) in 1990; and 3TC (lamivudine) in 1992. While none of these is superior to AZT, they can be used in patients who have become resistant to the effects of AZT.

With each drug, the enzyme reverse transcriptase attempts to incorporate the drug as its triphosphate into the growing strand of viral DNA, but because they all lack a 3'-hydoxyl group, they cannot form a 3',5'-phosphate linkage that is needed for DNA, and thus provide a termination point for the DNA strand and strand building ceases. After much early controversy, it was established that AZT did extend the lives of AIDS patients, and did slow down the progression from the HIV-positive state to that of full-blown AIDS. Therapy is now started when the patient's CD4 count has fallen to around 500, and the dose of AZT that produces a positive response can be as low as 0.5 gram per day.

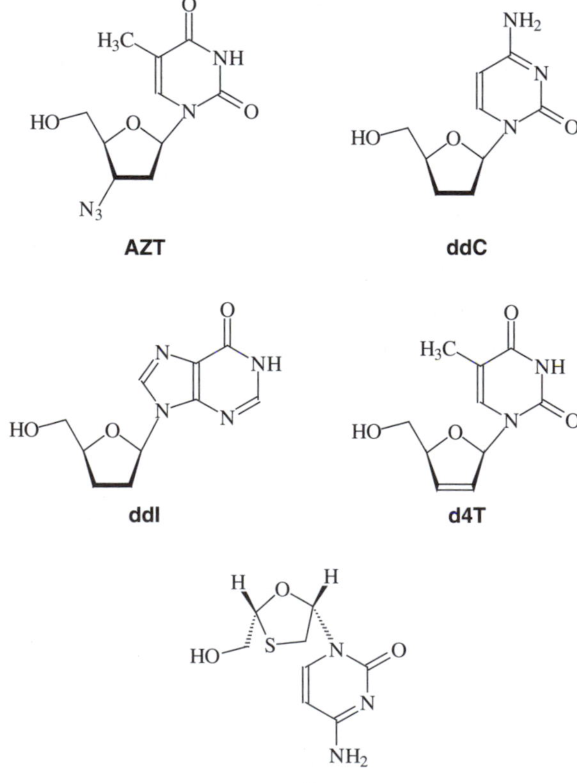

There are also now a number of non-nucleotide reverse transcriptase inhibitors and the most useful of these are nevirapine (Viramune, 1996) and delaviridine (Rescriptor, 1997), which appear to act at an allosteric site close to the active site of RT. This means that they block the subtle (and apparently essential) changes in the three-dimensional structure that occur when nucleotide substrates bind to the active site of RT.

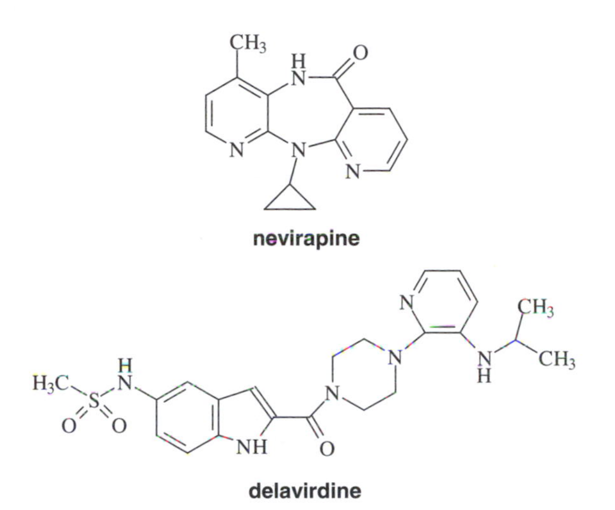

Inevitably, over time, HIV has mutated and has become resistant to all of the RT inhibitors. However, recent clinical studies have shown that early agressive treatment with two reverse transcriptase inhibitors, in conjunction with an HIV protease inhibitor, can reduce the body burden of virus to an undetectable level without the associated appearance of resistance. During the past 10 years, much effort has been expended on research with these protease inhibitors. Since the protease enzyme cleaves the first-formed large viral protein into the various structural proteins and enzymes of the mature virus, its inhibition results in a loss of infectivity. An enormous amount is now known about the structure of this enzyme from experiments where synthetic inhibitors were co-crystallised with the enzyme, thus providing valuable information about the three-dimensional structure of the enzyme's active site. These studies have revealed that the active site has two near-identical lobes with C2 symmetry, that is, rotation of one half of the structure through 180° would produce the other half of the structure. One of the main sequences recognised for cleavage by the enzyme is the tetrapeptide aparagine–phenyalanine–proline–isoleucine, and the generally accepted mechanism for cleavage involves twin aspartate residues as shown in Fig. 3.3. The rationally designed protease inhibitors all mimic the tetrahedral intermediate postulated for this cleavage but lack the cleavable peptide bond. This research benefited enormously from related work

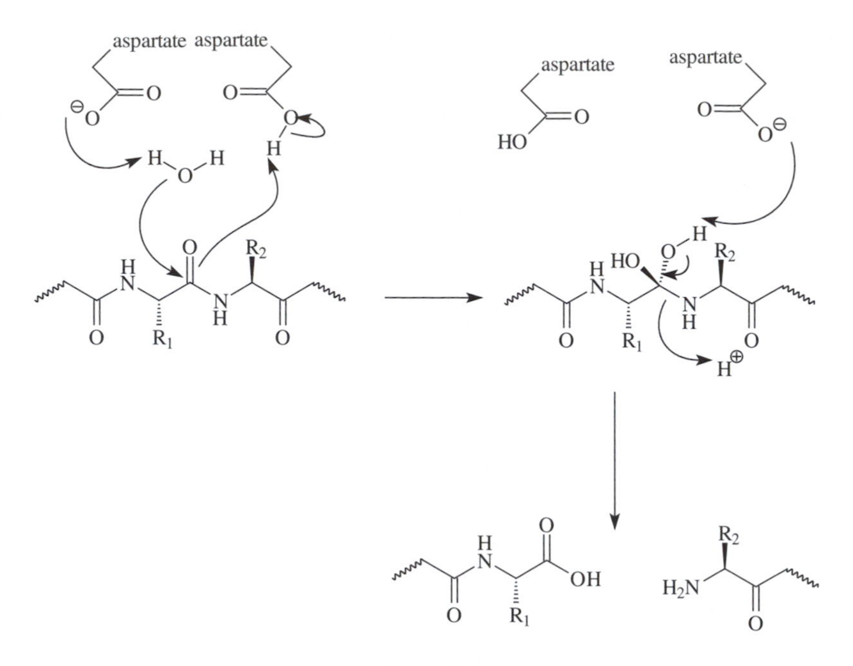

Figure 3.3

that had been carried out with other important aspartyl proteases like renin and the associated design and synthesis of renin inhibitors. A number of these HIV protease inhibitors are in clinical use, *e.g.*, saquinavir (Invirase), indinavir (Crixivan) and ritonavir (Norvir), although resistance soon appears if the drugs are used by themselves. These drugs are also relatively expensive to manufacture and are relatively poorly absorbed into the bloodstream.

There are several other areas where clinical candidates have appeared. These include inhibitors of the integrase enzyme like L-chicoric acid; viral fusion inhibitors like the complex polypeptide enfuvirtide (Fuzeon), which blocks HIV-1 infection through interaction with gp41; and various prodrugs of RT inhibitors like the phosphoramidate shown. There was also some initial interest in inhibitors of the host cell glycosidases. These enzymes trim the excess carbohydrate from the newly formed surface glycoprotein of the new virions. Two natural products, castanopermine from the Australian plant *Castanospermum australe* and deoxynojirimycin (DNJ) from various *Streptomyces, Bacillus subtilis* and mulberry species, have been the most intensively studied. An analogue of DNJ, *N*-butyl-DNJ, has shown considerable promise in clinical trials, as have a number of castanospermine analogues. All of these compounds, both natural and synthetic, probably function because they resemble the natural carbohydrate substrates of the glycosidases. Finally, budding of the virus from the host cell is suppressed by interferon-alpha, and there is some evidence that combinations of IFN-alpha and reverse transcriptase inhibitors may be beneficial. However, most of these remain of experimental interest rather than of proven clinical utility, and the costs of producing these newer drugs is also high. Enfuvirtide, for example, requires a 106-stage synthesis but is nonetheless being produced on the scale of three metric tonnes per year – its polypeptide sequence is based upon part of the structure of gp41.

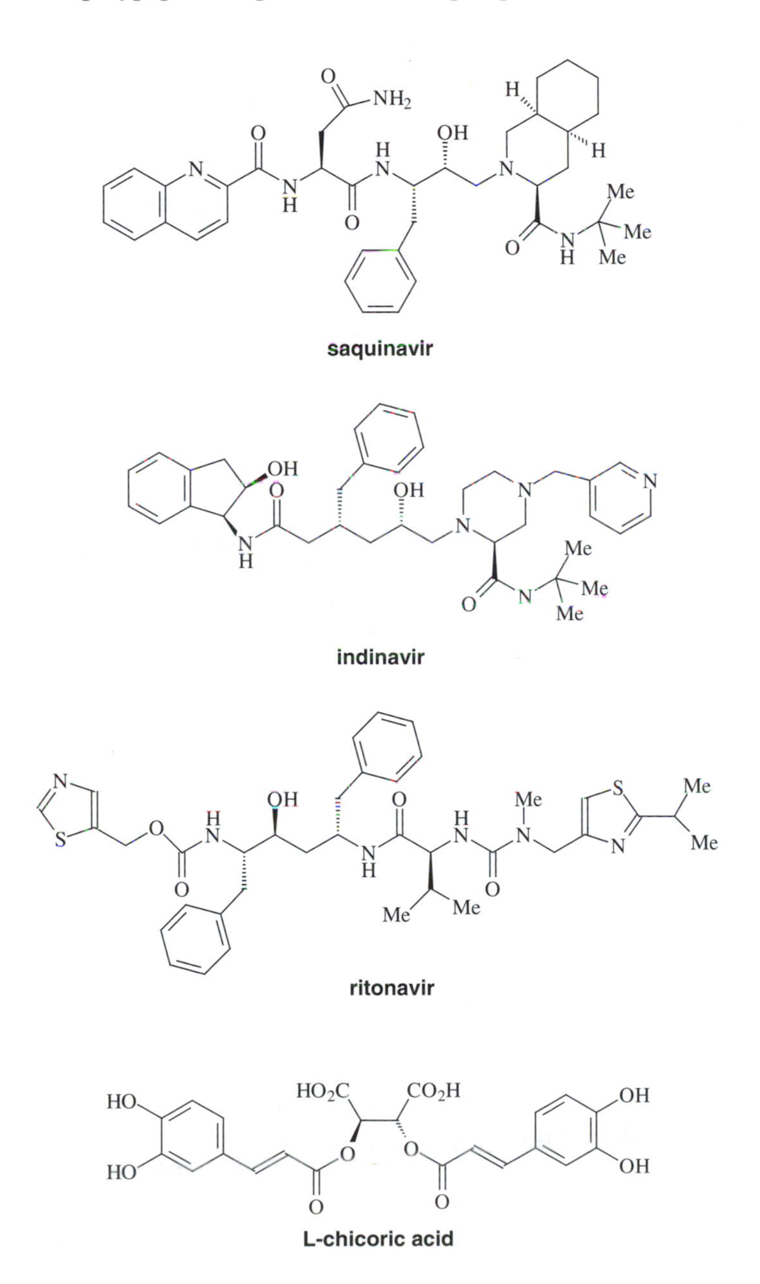

saquinavir

indinavir

ritonavir

L-chicoric acid

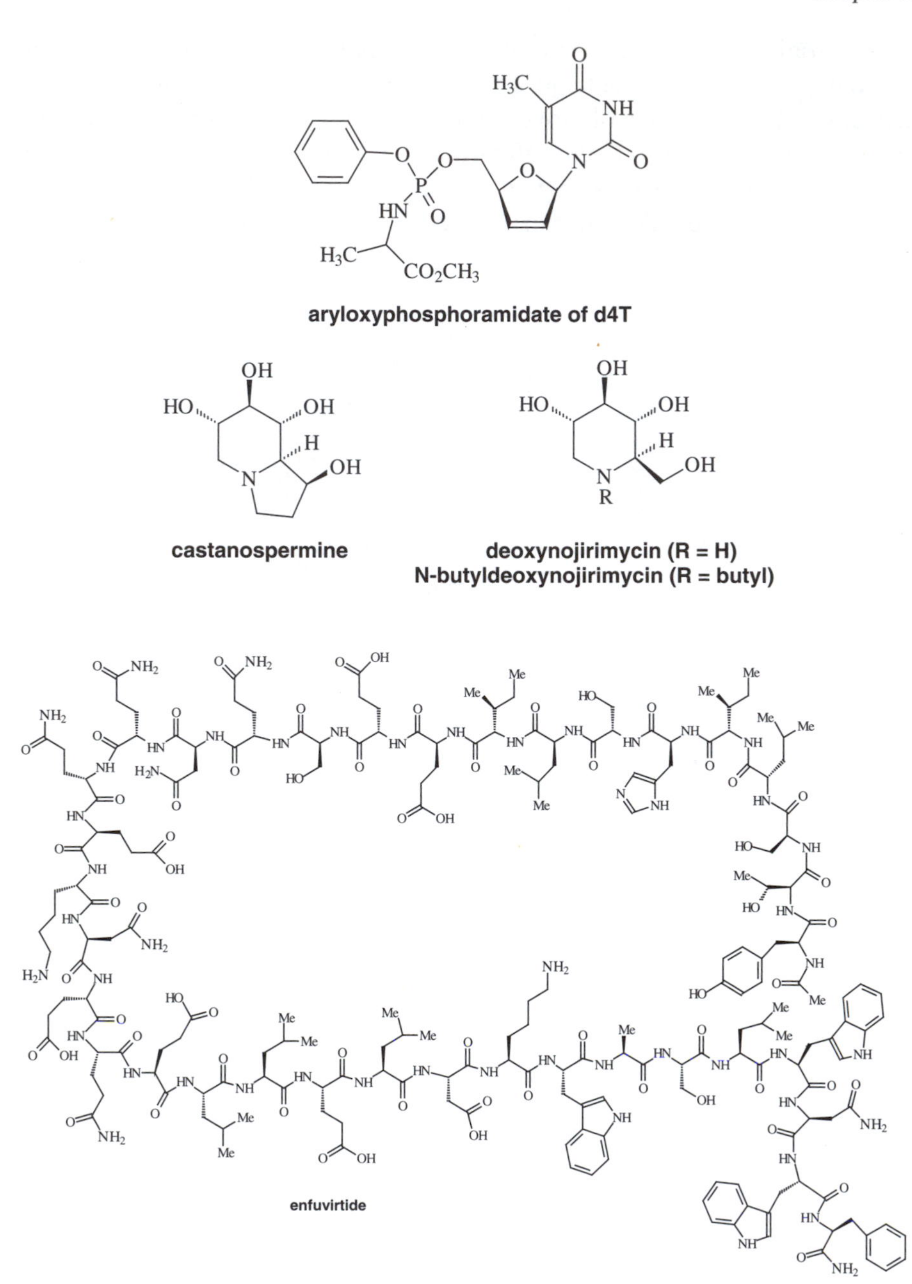

aryloxyphosphoramidate of d4T

castanospermine

deoxynojirimycin (R = H)
N-butyldeoxynojirimycin (R = butyl)

enfuvirtide

The overall effect of these various forms of chemotherapy, together with better methods of controlling opportunistic infections, has been to increase the length of survival after the onset of AIDS by about three-fold. Treatment of newly infected patients with the combination AZT, 3TC and the protease inhibitor Indinavir has reduced measured virus levels from a million RNA copies per millilitre of plasma to undetectable levels (around 200–400 copies per ml), and 90% of these patients remain in remission for more than a year. But there is a price! The cocktail of drugs needed for each patient costs around $10,000 to $15,000 per annum, and the annual market for HIV drugs now approaches $5000 million per year – and clearly these costs are far beyond the means of the poorer countries where HIV/AIDS is most prevalent and where an estimated 50 million people will die between now and 2020. Recently, several drug companies have offered to lower their prices for developing countries by as much as three-quarters, and this will go some way towards helping to reduce the anticipated death toll in these countries.

The future, however, looks a little brighter. The discovery of the co-receptors CCR5 and CXCR4 has opened up a new and rapidly changing area of research. These glycoproteins are the natural receptors for a range of so-called *chemokines*, which, once attached, attract a variety of other factors that mediate inflammatory reactions. Of particular interest is the fact that some Caucasians possess an altered form of CCR5, and these individuals exhibit marked resistance to infection by HIV. The possibility of designing drugs that will bind to CCR5 or CXCR4, or that will induce the production of chemokines to damage HIV-infected cells, is most alluring.

However, while we await the ultimate deterrant for HIV, it is sobering to consider how much more serious mankind's plight could have been if HIV was transmitted as efficiently as the influenza virus! We can take some comfort from the fact that (thus far!) it has 'chosen' the relatively inefficient sexual route of transmission.

HAEMORRHAGIC FEVER VIRUSES

While an AIDS plague is now unlikely (at least in the Western World), and was perhaps never a realistic possibility given the relatively inefficient route of infection by HIV, the haemorrhagic fever viruses have the potential to annihilate populations. Viruses like the Marburg virus, Ebola virus and Lassa fever virus are endemic in tropical regions and are highly contagious. They produce rapid and almost total destruction of white blood cells and their detritus clogs vital organs with resultant liver, kidney, lung and heart failure. In closed communities, the death rate is usually greater than 70% and can be close to 100%.

Historically, Marburg virus was the first to come to prominence. During August 1967, 21 workers at Behringwerke AG in Marburg, part of the German pharmaceutical company Hoechst AG, became ill with flu-like symptoms. These rapidly progressed to more serious virus-induced changes including enlarged spleens and lymph nodes, lowered white blood cell counts and severe loss of blood platelets with resultant fatal haemorrhaging. A doctor and a nurse at the hospital in Marburg also succumbed to the disease.

The clue to the source of this new disease was the contemporaneous appearance of the condition in Frankfurt, where four scientists at the Paul Ehrlich Institute, and a doctor and pathologist at the local hospital, who treated them, contracted the disease. But more significantly, a veterinarian and his wife in Belgrade, Yugoslavia were affected. This vet had examined a shipment of vervet monkeys that had arrived in Belgrade from Uganda. Half of the consignment had died in transit, but the remainder, and two further consignments, were shipped onwards to the laboratories in Marburg and Frankfurt. All of the German patients had either worked with these monkeys or had been infected by those who subsequently contracted the disease. Of the 31 persons infected, seven died and many of the survivors were incapacitated for years.

A research team in Marburg, Rudolf Siegert and Gustav Martini, were able to isolate and prove that the infective agent was a new virus, which they christened the Marburg virus. Later studies by the WHO showed that the virus was, not surprisingly, prevalent in Ugandan vervet monkeys, demonstrating once again the potential for transfer of disease from lower mammals to man.

Less than two years after these events, in early 1969, nurses at the mission hospital in Lassa, Nigeria, suddenly became ill with high fever, rashes, heartbeat irregularities and ultimately bodywide haemorrhage with almost every organ clogged with detritus from damaged white blood cells. The disease was clearly highly contagious, because almost everyone who came into contact with an infected patient subsequently contracted the condition. One of the nurses, Lily Pinneo, who was dangerously ill with the disease, was flown to the USA where she eventually recovered following two months of intensive nursing. Samples of her tissues and those of her dead colleagues were studied at Yale University by Jordi Casals, who had been intimately involved with the identification of the Marburg virus. Although he was working under the most stringent containment conditions, he too contracted the disease in June 1969. As his condition deteriorated, the decision was taken to administer a sample of Pinneo's plasma in the hope that it contained antibodies. It obviously did, because Casals made a rapid recovery.

The organism responsible for the disease was eventually shown to be a virus very similar to various known viruses carried by rodents and bats. It was apparently transmitted through the air (as viral particles) and from contaminated urine and blood. Several outbreaks occurred in subsequent years in Nigeria, Liberia and Sierra Leone, and the source of these infections was eventually traced to the local rat population and to one species in particular, *Mastomys natalensis*. This investigation had involved the trapping of a large number (641 in all) of small vertebrates in the vicinity of the outbreaks and then examining them for the presence of the virus. The results were clear-cut. The virus was not present in the 141 samples of *Mus musculus* (housemouse) studied, nor in the 50 black rats (*Rattus rattus*), but was present in 14 out of 82 samples of *Mastomys natalensis.* Due to a partial eradication of the competing species, the black rat, the smaller *M. natalensis* had become much more populous and had begun to live in close proximity to the human populations. The virus was probably initially transmitted via food contaminated with rat faeces.

In August 1976, probably the most deadly virus of them all first appeared in a mission hospital in Yambuku, Zaire. A number of patients developed flu-like symptoms, which rapidly progressed to a haemorrhagic fever and within six weeks, the patients, 38 of their relatives and almost all of the hospital staff had succumbed to the disease. The Center for Disease Control in Atlanta was asked for help, and a massive operation began to control the spread of the disease and to identify its source. Five laboratories, in Belgium, France, Germany, the USA and one at the Microbiological Research Station at Porton Down in the UK, were charged with identification of the virus. During the course of this work at Porton, one of the scientists, Geoffrey Platt, pricked his thumb with the needle of a syringe containing the virus. Although there was no evident wound and no bleeding, six days later he became ill with a fever, nausea and abdominal pains. He was immediately treated with human interferon and continued to receive injections for the next two weeks. Despite this prompt action, he suffered a crisis on day four with a high fever associated with severe malaise, weakness, profuse watery diarrhoea and persistent vomiting. A rash covered all parts of his body and he also suffered from severe thrush. After two days in this state, he made a gradual recovery. It is likely that the prompt administration of interferon significantly changed the course of the infection, but it provided a cogent example of both the infectivity of the agent and its potential to cause a life-threatening condition.

The virus was eventually proved to be Marburg-like, and was worm-like in shape with a coil at one end, but was much more pathogenic than the Marburg virus. The new virus was named after the river Ebola that ran through Yambuku. These viruses are members of the family known as *filoviruses*. Separate outbreaks of Ebola virus occurred near Maridi and N'zara in Sudan, with 284 cases and 151 deaths, which represented a somewhat better survival rate from that experienced around Yambuku where there were 318 cases and a mortality rate of around 90% of those infected. This high mortality was undoubtedly due to the use and reuse of unsterilised syringe needles, by the mission medical staff. More recently, in May 1995, there was another major outbreak, in Kikwit, Zaire, and to date, an animal reservoir for the Ebola virus has not been identified.

Clearly, this is a difficult virus to study due to the remote African locations and the severe risk to the researchers. However, serum samples were collected from patients during two outbreaks in the Gabon in 1996, and several key features were identified. In survivors, there was a rapid increase in the levels of the immunoglobulin class of antibodies (humoral response) directed against the nucleoprotein of the virus, and also activation of cytotoxic T-lymphocytes. In non-survivors, the humoral response was weak, although they did produce a high level of γ-interferon. This latter inverse relation between γ-interferon levels and survival may at least be a useful prediction of those who are at most risk from the infection.

Finally, the outbreak of infections caused by a hantavirus amongst the Navajo population in New Mexico in 1993 provides an excellent example of the effects of a climate change on disease emergence. The symptoms of this disease included fever, headache and then rapidly progressing respiratory distress, and death within a few days. Infection with hantaviruses was not new, having been first observed during the Korean War (1951-1954). More than 2500 American GI's suffered flu-like symptoms followed by kidney failure, and there were over 120 deaths. The viruses are commonly found in animals and some clever detective work revealed that the Navajo outbreak originated from the faeces of the deer mouse, *Peromyscus maniculatus*. The mouse had moved into human habitations during a period of prolonged drought. This had been followed by a winter of record snowfalls and a damp spring, and the resultant lush vegetation in and around the habitations had led to an explosion in the mouse population another timely reminder that we humans live in a rather shaky equilibrium with other species and with our climate.

Outbreaks of hantavirus infection in Europe have also occurred, and this has usually been due to human encroachment on the territories of wild animal species. In North-Eastern France, for example, there have been at least 500 cases of infection during the past 20 years. This was due to inhalation of wood dust during tree-felling, in areas where the red bank vole was a common species. The virus was present in the urine and faeces of these animals, and these were absorbed on the trees that were felled.

The westwards migration of yellow fever in the 17th century, is perhaps the best example of the emergence of a viral disease, triggered by man's disturbance of a native environment. The virus has probably always been endemic in West Africa, and is usually confined to the monkey population. The mosquito, *Aedes aegypti,* transmits the virus to human populations that come into close proximity with monkeys. In Africa, most of the native population is immune to the virus, having encountered it in their youth and then survived with immunity or perished from the infection. When the slave-traders reached West Africa in the mid-17th century, they unwittingly carried the mosquito along with their human cargo to the American and European ports. Devastating outbreaks of 'Yellow Jack' decimated ships' companies, and the first well-documented outbreak in the western hemisphere occurred in the Yucatan Peninsular, Mexico, in 1648. The disease soon reached most of the Americas, and the European seaports. In one major epidemic in 1793, 15% of the population of Philadelphia died from the disease, while New Orleans experienced 33 epidemics between 1800 and 1850.

The mosquito is not very hardy, and will only thrive in a humid environment at a temperature in excess of 25 °C. It must live close to human habitation, and breeds especially well in the gutters of houses, and in other domestic water supplies. After being bitten by an infected mosquito, the victim remains well for a while, but after 3–6 days suffers flu-like symptoms, with headache, backache, fever, nausea and vomiting. This is followed by a brief respite before the main stage of the illness, during which there is serious jaundice and haemorrhaging (the Spanish word for the condition is *vomito negro*), as the heart, kidneys, and most particularly, the liver suffer severe (and usually irreversible) damage, in short, the classic symptoms of a haemorrhagic fever. Mortality ranges from 10 to 90%. Those who survive never suffer from a recurrence of the disease.

Progress towards a treatment was very slow, since the mode of transmission was not established until 1901. A Cuban physician, Carlos Finlay, had speculated in 1881 that the vector might be a mosquito, and this was confirmed in the early years of the 20th century. The Americans Walter Reed and James Carroll and their colleagues, succeeded in transmitting yellow fever between human volunteers via the bite of an infected mosquito. Two groups of soldiers were exposed either to the clothing and bedding of yellow fever victims (but no mosquitoes), or to the bite of mosquitoes who had taken blood from yellow fever victims. The latter group contracted yellow fever, while the former group remained healthy.

The virus was not, however, isolated until 1928, and a vaccine was first developed by the South African Max Theiler in 1939. Early attempts to eradicate the mosquito by physical means, and later through the use of insecticides like DDT, were only partially successful, and the disease is still endemic in much of West Africa and in South America, but not, surpisingly, in Asia. It has been suggested that this may be due to the fact that it closely resembles the virus that causes dengue fever, which is endemic in all of Asia. They are both *arboviruses*, which stands for arthropod-borne viruses, are both transmitted via the bite of a mosquito and cause a haemorrhagic illness. Survivors of dengue fever would be expected to have partial immunity (at least) to yellow fever.

Rift Valley fever is also transmitted by mosquitoes of the *Aedes* family, and an epidemic, in the early 1970s, in Egypt affected more than 200,000 people, with 600 deaths. On this occasion, the precipitating event was the construction of the Aswan dam, with the associated increase in water available for mosquito-breeding areas.

These haemorrhagic fever viruses are presently confined primarily to tropical regions. However, the probability of worldwide climate change due to global warming coupled with the encroachment of man on the remaining wild places of the planet provide the necessary scenario for the emergence of new viruses and possible epidemics. This situation is exacerbated further by the capacity for rapid mutation that these viruses possess. They are RNA-(−)-viruses. This means that the viral RNA is not in the correct form to act as a blueprint for viral DNA production. It must first be used as a template for the production of a complementary RNA-(+)-strand. This is effected by a viral RNA polymerase, and these enzymes are notorious for the number of mistakes that they make. Mutation is thus a frequent event.

Despite the dangers posed by these viruses, a much greater concern for western societies is the ever-present spectre of cancer, and this is the subject of the next chapter.

Chapter 4

Cancer: The Disease and its Treatment

Being a surgeon in ancient times was a precarious existence: if the patient survived, all was well, but if the patient died, the surgeon could easily find himself buried alongside the deceased. Nonetheless, the Ebers papyrus (the ancient Egyptian treatise on the medicinal uses of plant and animal extracts and dating from about 1500 BC) provides evidence that Egyptian surgeons regularly identified and excised tumours: 'If a growth comes-and-goes under thy fingers (probably benign) treat it with the knife.' A number of tomb paintings clearly show evidence of human tumours. The danger of a cancer spreading (*metastasis*) following incomplete surgical removal was noted by Hippocrates in his *Aphorisms*: 'It is better not to apply any treatment in cases of occult cancer, for if treated the patients die quickly; but if not treated, they hold out for a long time.' And in a similar vein, the Roman physician Celsus recommended surgery for early diagnosed tumours, claiming that late intervention aggravated the situation.

As usual, the great Greek medical writer Galen provided precise instructions in his books entitled *The Therapeutic Method*:

> When a tumour has risen to a notable size, there is no cure without surgery. The goal is to proceed with a round incision so that the entire tumour can be excised... . A cure will be more readily effected when the black humour is not so heavy, since purgative action of drugs will easier lead to a cure.

The reference to the 'black humour' accords with his belief that all ailments were due to an imbalance in the four body humours: blood, phlegm, black bile and yellow bile. Galen was also one of the first to describe the appearance of a large cancer (from the Latin *cancer* via the Greek *karkinos*=crab) with its characteristic inflamed 'crab-like' projections showing the radial spread of disease to the neighbouring lymph nodes. Galen also observed that "cancerous tumours develop with greatest frequency in the breast of women...such tumours have their source in the black bile, a superfluous residue of the body", and he believed that the breasts were most frequently affected because they were not able to divest themselves of this noxious residue.

These observations have to be noted in the context of the disease patterns in ancient times. Cancer was undoubtedly comparatively rare since the average life expectancy was quite short – probably 30–40 years – and cancer is usually a disease of middle to old age. Surgeons and physicians were much more concerned with the treatment of gastrointestinal problems caused through consumption of contaminated food or water; sexually transmitted diseases; and battlefield injuries.

The true causes of cancer were completely unknown, and this ignorance lasted well into the 19th century. The first serious suggestions about the origins (aetiology) of cancer appeared in 1700 when the Italian physician Bernardino Ramazzini in his book *De Morbis Artificium Diatriba* commented upon the prevalence of breast cancer amongst nuns, especially in contrast to the relative rarity of the condition amongst women who had borne and suckled many children. He had no way of knowing the importance of female hormones in these relationships. It is of interest that Ramazzini also drew attention to the frequency of pneumoconiosis in miners and stone-masons owing to their inhalation of fine dust particles. The English physician John Hill commented in 1761 about the predisposition of snuff-takers to contract nasal cancer, probably the first indication that a tobacco product was harmful to health; but it was Percivall Pott who laid the foundations for modern cancer epidemiology.

Percivall Pott was born in London in 1714 and became assistant surgeon at St. Bartholomew's Hospital in 1745, rising to the position of Chief Surgeon in 1749. He became a specialist in spinal diseases and was the first to recognise that a form of tuberculosis (*i.e.*, an infection with the microorganism *Mycobacterium tuberculosis*) was responsible for a damaging inflammation of the spinal vertebrae, leading ultimately to spinal curvature and paraplegia. The condition, now treatable with antibiotics or corrective surgery, still bears his name. He also wrote several learned books, including *A Treatise on Ruptures* in 1756; but it was his observation in 1775 that most men who contracted scrotal cancer had worked in their youth as chimney sweeps that provided the first real link between an environmental carcinogen (soot) and a cancer.

The success of the Industrial Revolution had been largely due to the cheapness and ready availability of coal, and the large houses of the many successful businessmen (and the landed gentry) also used coal for all cooking and heating. These houses had huge chimneys that required frequent cleaning, and hundreds of small, scantily clad (or naked) boys were employed to scramble around inside them with brushes. Inevitably, their skin was exposed to the chemical compounds present in the soot over a period of many years, since the soot was most easily trapped in the folds of skin and resisted the limited attempts to remove it by washing. In later life, many of them succumbed to scrotal cancer.

Despite Pott's well-publicised warnings about the risks, the scandalous use of boy chimney sweeps continued until the late 1800s. An end to the practice started with the Chimney Sweepers' Act in 1864, which was rushed through parliament as a response to the public outcry caused by the publication of Charles Kingsley's book *The Water Babies* in 1863. This provided a graphic though fictional account of the relations between Tom the boy chimney sweep and his master Grimes; hence, the public response was more humanitarian than a response to fears about cancer. The Act was rendered useless by a powerful lobby of rich householders, local authorities and other objectors, and it took a further Act, introduced by Lord Shaftesbury in 1875, to make the practice illegal.

Other instances of cancer produced by exposure to chemicals in the workplace soon emerged. Von Volkmann described in 1875 the high incidence of skin tumours amongst workers in the tar and paraffin industries in Halle, a suburb of Manchester. Joseph Bell, a physician in Edinburgh, described in 1876 a 'paraffin cancer' amongst Scottish workers in the oil shale industry; and in 1887, 'mule-spinners' cancer' was ascribed to the use of mineral oil by those working in the Lancashire cotton-spinning industry.

AETIOLOGY

Actual proof of the link between chemicals and cancer had to await the experiments of Yamagiwa and Ichikawa in 1915, who produced malignant epithelial tumours by application of coal tar to the ears of rabbits. [The epithelium is a collective term for the large variety of cell types that provides the lining of the mouth, nose, intestines, respiratory tract and skin.] In a similar vein, Passey painted the skin of mice with an extract of soot and produced malignant growths. But isolation of the first actual carcinogen – benzo[*a*]pyrene or simply benzpyrene – and chemical synthesis of a pure carcinogen – dibenzanthracene – both happened in the early 1930s. These polycyclic aromatic compounds are common constituents of soot, tars (including those from tobacco smoke), car exhausts (especially from diesel engines) and (in trace amounts) in burnt food. It is interesting to note that it has taken a further 70 years to untangle the actual molecular mechanism by which benzpyrene in cigarette smoke induces a normal cell to become a cancer cell (Fig. 4.1). The words of the French physician Lebert in his *Traite pratique des maladies cancereuses* (1851) are as pertinent now as then: "Il faut que, pour un pareil travail, un homme profondement verse dans la connaissance anatomique et pathologique du cancer se réunisse a un chimiste qui, de son côté, soit au courant des progrès les plus récent de la science." Or in essence: "To make progress in cancer research, a biologist skilled in the anatomy and pathology of cancer should collaborate with a chemist who is

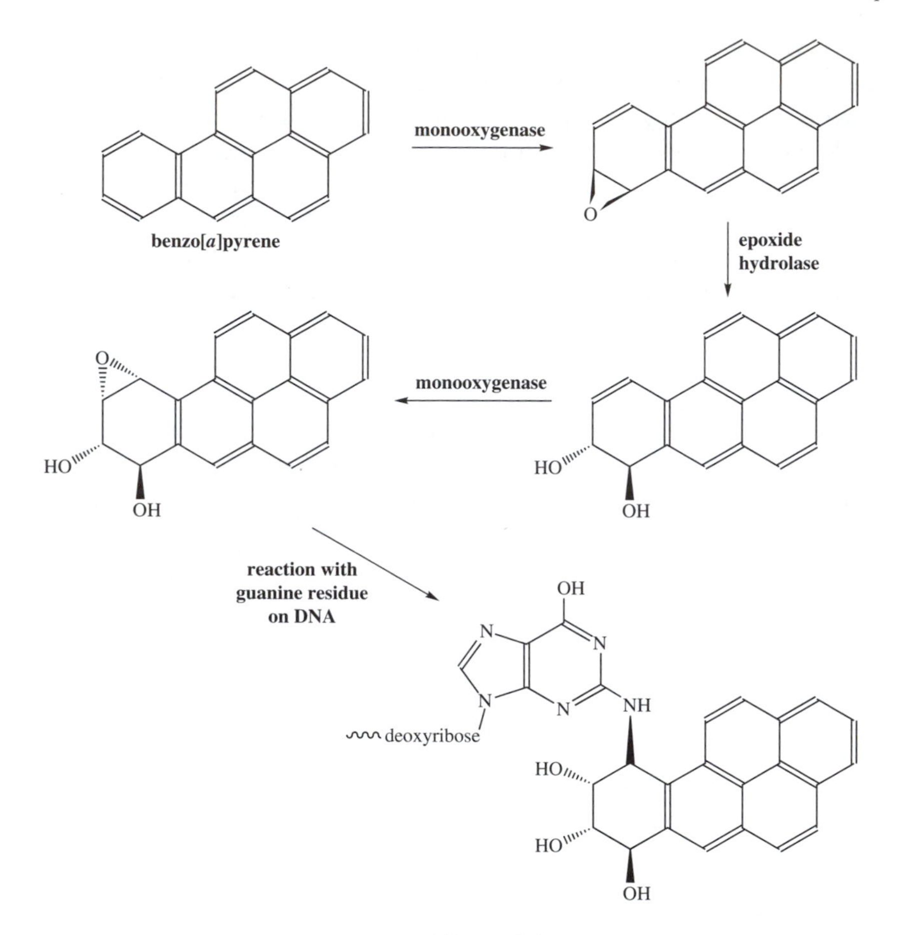

Figure 4.1

fully conversant with all of the latest advances in his field." This must be one of the earliest pleas for biologists to collaborate with chemists.

It would be a gross oversimplification to ascribe all cancers to chemicals in our environment. The figures drawn from many patient surveys over the past 50 years show quite clearly that 30–35% of all cancers (most commonly of the lung, bladder, kidney and pancreas) are tobacco-related, while around 40% are probably food-related. A diet high in fats, low in fibre and with a predominance of grilled and fried foods predisposes the consumer to cancer of the large bowel, breast, pancreas, prostate, ovary and endometrium, representing perhaps 30–35% of all cancers. Of less significance are diets high in nitrite or nitrates and low in vitamin C, which appear to have a link with stomach cancer (2–3% of all cancers), and the excessive consumption of alcohol or exposure to mycotoxins (toxic natural products produced by microorganisms in certain foods), which can cause liver cancer (3–4% of all cancers). Other cancers are probably caused by exposure to natural radon gas, which emanates from granite, to ionising radiation (X-rays, γ-rays), sunlight or to certain viruses, or are due to an unfortunate genetic disposition.

Before considering how cancers arise and the possible strategies that may be used to treat them, it is instructive to consider these potential environmental carcinogens in more detail. The one irrefutable fact is that carcinogens present in tobacco smoke cause the vast majority of cancers of the lung but also a large number of cancers of the mouth, pharynx, larynx, oesophagus, bladder, pancreas and kidney. Lung cancer is responsible for more male cancer deaths (about 20,000 per year in the UK – Cancer Research UK figures for 2002) than any other cancer, and for women, is just greater than those caused by breast cancer (each around 13,000 per year). By the age of 60, smokers have a 10 times greater chance of dying from lung cancer than do life-long non-smokers. If one adds the increased risks of heart attack and stroke amongst smokers, the dangers of the habit referred to by James I as – "loathesome to the eye, hateful to the nose, harmful to the brain and dangerous to the lungs" – are manifest. In total, tobacco smoking causes about three million deaths worldwide, and if sales continue to grow in the developing world, a figure of 10 million deaths per annum may be reached by 2030! The morality of this trade is now in some doubt given the recent admission by three American tobacco companies that smoking is harmful to health. In a historic decision on June 19th, 1997, after decades of legal wrangling, the US tobacco companies Philip Morris and R. J. R. Reynolds, and the British company British American Tobacco, agreed to pay $360 billion primarily to victims of smoking-related diseases, but also to fund an anti-smoking campaign. This punitive fine of $360 billion represented 'punishment' for past sins in the USA, but other court cases are pending in several countries including the UK.

Smokers cannot claim that they have not been warned, since cogent evidence was published as long ago as 1954 by the British epidemiologists Richard Doll and Richard Peto who observed a clear link between the incidence of lung cancer and smoking. They followed the fates of 40,000 doctors in the UK over 20 years, and this seminal report was followed in 1981 by their more general summary of what they termed 'avoidable cancers', tobacco-related disease being the most obvious form. What made this report such a landmark was the accumulated data and evidence that they cited in support of the supposition that certain (probably most) cancers were related to lifestyle or environment. For example, the fact that skin cancer is over 200 times more prevalent amongst the fair-skinned residents of Queensland, Australia, than amongst the darker-skinned inhabitants of Bombay, India, provides cogent evidence of the effects of ultraviolet light on an immigrant population whose skin colour was more designed for the relatively sunless

Highlands of Scotland than the tropical climate of Queensland. The very high levels of liver cancer (as much as 100 times higher than in Britain) in parts of Africa (notably Mozambique) have been ascribed to the presence of mycotoxins (aflatoxins) in the groundnuts that form a major part of the diet of people living in these regions, although an association with infection with the hepatitis B virus is also very likely. Similarly, the very localised incidence of oesophageal cancer in parts of South Africa, Iran and China has been connected with particular peculiarities in the local diets. In Normandy, the high local incidence of this same cancer is ascribed to the consumption of the strong local apple brandy called Calvados, although the identity of the carcinogens is not known.

The importance of Doll and Peto's report cannot be overemphasised. In addition to their revelations and predictions about the environmental aspects of cancer epidemiology, they also made highly perceptive predictions about other potential environmental problems. For example, their predictions about the effects of chlorofluorocarbons (CFCs) in the atmosphere (a restatement of comments made in a 1975 report) leave one in little doubt that politicians and industrialists deliberately ignored warnings for well over a decade before taking steps to avoid an environmental catastrophe. Doll and Peto wrote: "The CFC's persist in the atmosphere where they will react with ozone, reduce its concentration, and hence permit more UV light to reach the surface of the earth. If this happens, the incidence of skin cancer, including the relatively fatal melanoma of the skin, must be expected to rise."

Genetic make-up is certainly a factor that predisposes an individual to cancer – a familial pattern of breast cancer is, for example, well-established. However, an additive and even predominant environmental effect is clearly evident when one studies migrant populations. In Japan, the incidence of stomach cancer is high, probably due to a high consumption of salted fish and pickled foods; but the incidence of prostate cancer is low, perhaps because this has an association with the consumption of large amounts of red meat. However, in migrant Japanese communities in the USA, the incidence of these two cancers more closely mirrors that of the native US population, suggesting that the adopted US diet has affected their predisposition to the two cancers. In addition, black Americans have similar incidences of the main cancers to white Americans, but these differ markedly from those seen in black Africans.

Increased cancer risk caused by carcinogens in the work-place is also well-established. In addition to the effects of benzpyrene and other polycyclic aromatic compounds already mentioned, the high incidence of mesothelioma, a rare lung cancer amongst those exposed to the blue asbestos fibres is irrefutable. There is also good evidence that other cancers have been caused by such diverse chemicals as vinyl chloride (used in the manufacture of PVC), and certain azo dyes used as food colorants. One particularly bizarre example of the latter class was the use of the yellow dye chrysoidine in maggots used by fishermen. The anglers would warm the maggots in their mouths before baiting their hooks and there was a spate of bladder cancer amongst anglers in the 1980s, ascribed to the consumption of chrysoidine. The subsequent proven carcinogenicity of this dye, along with the relative youth of those affected with bladder cancer, provided evidence firm enough to ban the use of the dye in food products.

More insidiously, our own sex hormones, and the synthetic analogues present in the oral contraceptive and in the drugs used for hormone replacement therapy (HRT) can predispose us to developing certain tumours. For example, prostate cancer is undoubtedly caused by the effects of a lifetime's exposure of this small gland to the male hormone testosterone and its metabolic product dihydrotestosterone. The effects of long-term exposure to the female hormones (oestrogens and progesterone) are clearly evident in the increased incidence of breast cancer in women who experienced an early menarche (commencement of periods) and a late menopause. Long-term use of the contraceptive pill and of hormone replacement therapy (HRT) also increases (very slightly) the risk of breast cancer.

Viruses have been implicated in the development of at least two cancers. Infection with the human papilloma virus is a major risk factor for cervical cancer, and like any sexually transmitted disease, the cancer is more common amongst those who have had numerous partners and is almost unknown in nuns. But the most well-established link is between Burkitt's lymphoma and the Epstein-Barr virus (one of the herpes viruses) in African countries. In developed countries, this virus causes glandular fever (infectious mononucleosis), and the different progress of the disease in Africans is apparently due to an immune system that is usually suppressed through a long-term assault by malaria parasites like *Plasmodium falciparum.*

To understand how these various agents can cause or predispose the individual to tumour induction, we need to consider how the growth and differentiation of normal cells takes place, and how these processes can be disrupted. An adult human is constructed from about 30 million million cells (30×10^{12}), which have diameters ranging from about 5 μm(like a red blood cell) to 120 μm in diameter (like the ovum), but typically are in the range 10–20 μm. These cells have three main components: a membrane that completely encloses the cell, a nucleus that contains the chromosomes and their constituent genes (the 'blueprint' for all life processes) and the cytoplasm in which most of the biochemical processes take place. The cell membrane is constructed from two layers of lipid (mostly fatty acids, both saturated and unsaturated) that has glycoproteins embedded in it. These glycoproteins are intimately involved in receiving and modulating the chemical signals that impinge upon the cell, and ensuring that they pass into the cell and onwards to the nucleus – they are the *receptors* for drugs, growth factors, *etc*. All kinds of signalling strategies are used in the cytoplasm, but most involve raising or lowering of concentrations of metal ions (*e.g.*, calcium), or activation and deactivation of certain proteins (*e.g.*, kinases or phosphatases). These are enzymes that catalyse the addition or removal, respectively, of a phosphate group to or from another protein or enzyme. In addition, the steroid hormones and similar agents attach themselves to receptors in the cytoplasm, and these hormone–receptor complexes then pass into the nucleus to convey their message (Fig. 4.2).

Ultimately, the signals impinge upon the DNA of the genes in the nucleus and trigger a burst of activity. The DNA contains the blueprint that determines most of our physical and probably personal characteristics, and the mechanisms by which this invisible encoded information is translated into our very visible features and other characteristics are shown in Fig. 4.3. The essence of this amazingly complex, but almost unerringly accurate, translation process is the 'reading' of the genetic message encoded in the messenger RNA (m-RNA – itself obtained by transcription from DNA) as it threads its way through the ribosomes. These function like a factory production line in which the m-RNA orders the assembly of a discrete sequence of amino acids into a growing protein. It is these proteins that provide most of our physical structure and (as enzymes) control the biochemical reactions that give us life with its rich array of physical and mental attributes. In addition to the continuous translation of genetic information into biochemical reality, most cells require their DNA to be duplicated periodically and for the cell to

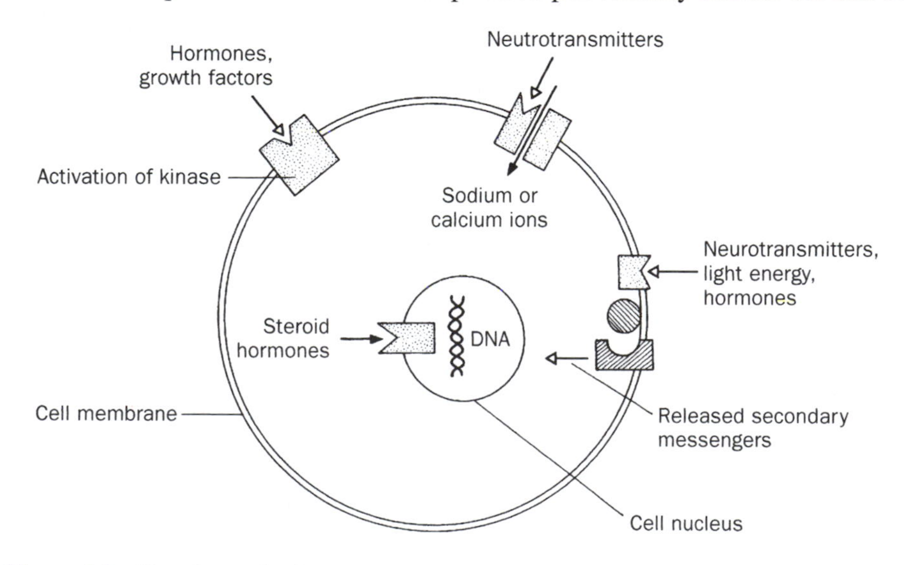

Figure 4.2 *Signal transduction*

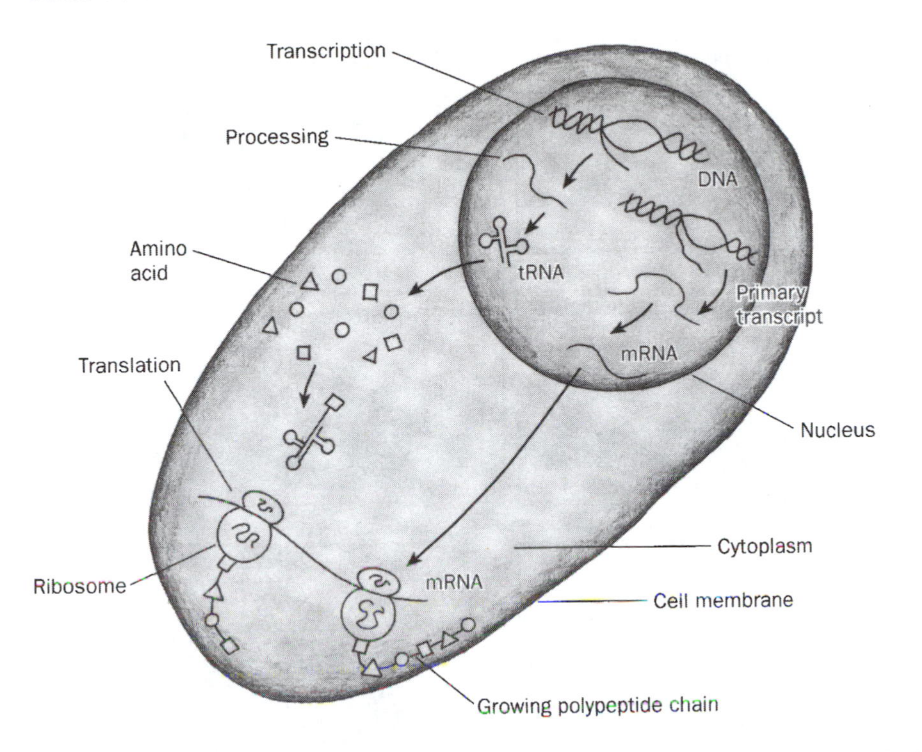

Figure 4.3 *DNA transcription and the biosynthesis of proteins. The process by which genetic information in the form of DNA is used as a blueprint for the production of proteins and enzymes has three stages. First, the twin strands of DNA (the double helix) become partially unwound and – with the assistance of the enzyme RNA polymerase – the 'genetic message' is used as a code for the production of RNA. This is transcription. Second, the RNA is processed into a variety of smaller RNA species: transfer RNA (tRNA), messenger RNA (mRNA), and ribosomal RNA (rRNA). These pass out of the nucleus and into the cytoplasm. Here the various kinds of tRNA take up discrete amino acids. These tRNA – amino acid conjugates can recognize signals on mRNA that are specific for the particular type of amino acid. Therefore, as the mRNA threads its way through the ribosome (a complex structure comprising rRNA and a number of different proteins), amino acids are joined together in a sequence, thus generating a growing polypeptide chain. This ultimately yields the fully formed protein or enzyme. This is translation*

reproduce itself, that is, to divide and produce two new daughter cells – the process known as *mitosis*.

Clearly, there are numerous points during the translation and duplication processes where mistakes can occur or where the delicate control mechanisms can break down, thus initiating a train of events that turns a normal cell into a rogue cell and thence (perhaps) into a cancer cell. Most normal adult cells only undergo mitosis to replace dead or damaged members of their class. This can be seen most easily during the healing of a wound. The damage is soon repaired as new cells are produced in response to signals released at the site of the wound; but the process of replacement ceases once the damage has

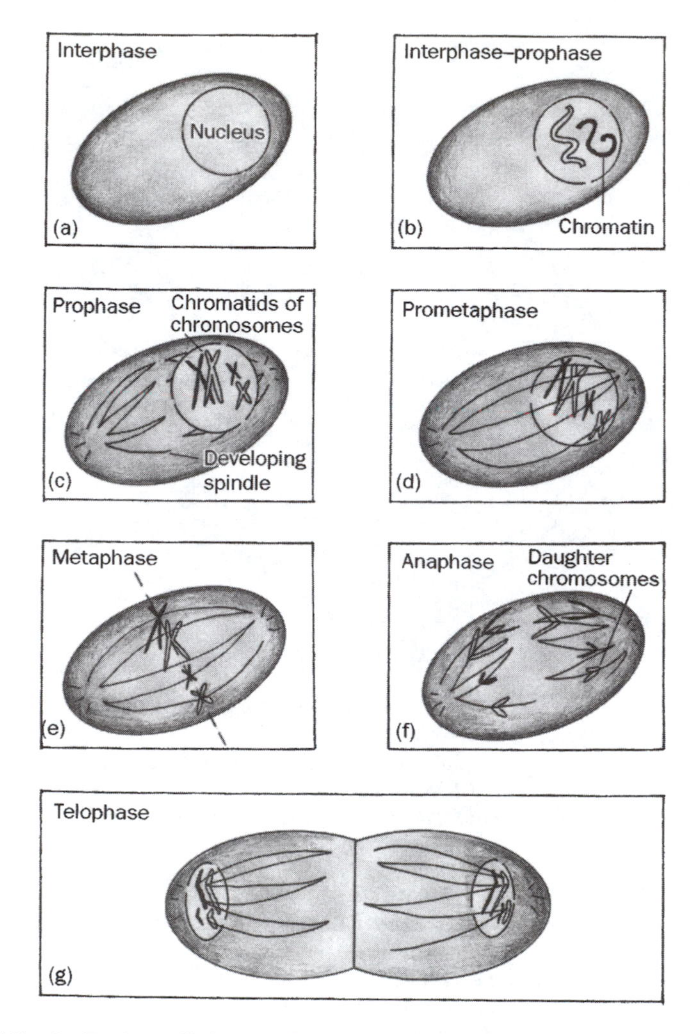

Stages in Mitosis: During cell division the genetic material of the cell (the genome) is duplicated and then distributed as identical sets of chromosomes to the developing daughter cells. The chromosomes condense during ***prophase*** *to form pairs of identical* ***chromatids*** *of DNA coiled together. At* ***metaphase*** *these chromatids are assembled at the centre of the cells by the developing microtubules. During* ***anaphase*** *the chromatids seperate and each daughter chromosome moves towards what will become the nucleus of the developing daughter cell. Finally during* ***telophase*** *the nuclear envelope forms around the chromosomes and the cell wall grows to envelope these nuclei forming two new cells*

been repaired. An understanding of the *cell cycle* and its control is clearly the key to the design of anti-cancer drugs. The cycle comprises four distinct stages (Fig. 4.4):

- G_1 phase (gap 1 phase) during which the cell increases in size through the synthesis of proteins and prepares for duplication of its DNA.

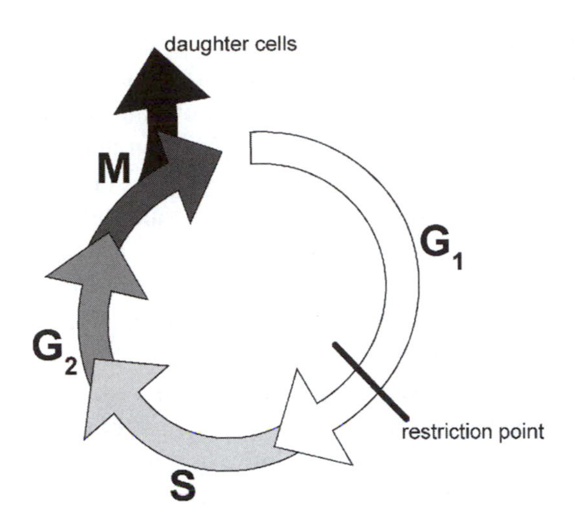

Figure 4.4 *Cell cycle*

- S phase (synthesis phase) when the DNA replicates to produce two sets of chromosomes.
- G_2 phase (gap two) during which the cell prepares for mitosis.
- M phase (mitosis) when the cell division occurs to produce two daughter cells each with a complete set of chromosomes.

This entire cell cycle 'clock' is controlled by a number of proteins called *cyclins* and their associated cyclin-dependent kinases (CDKs). These are associated with one another to control entry into and progression through the various stages of the cell cycle. For example, there is a key restriction point (see Fig. 4.4) during G_1, when a growth-promoting signal molecule will attach to cyclin D, which is associated with CDK 4 or CDK 6. The resultant complex then interacts with the growth-inhibiting protein called Rb – the so-called 'master brake' (Rb is an abbreviation for retinoblastoma since mutations in the gene that produces the Rb protein are implicated in this type of tumour), and this is associated with inactive forms of the transcription factors that will initiate replication of DNA. The interaction with the cyclin D/CDK 4/6 complex triggers a transfer of several phosphate groups (from adenosine triphosphate, ATP) onto Rb, and this results in the release of active transcription factors. These then activate the genes to replicate and code for the synthesis of the requisite proteins and enzymes (cyclins E and A, DNA polymerase, thymidine kinase, dihydrofolate reductase, *etc.*) needed for progression into the synthesis phase. It is not difficult to appreciate how an imbalance in the levels of the various cyclins or CDKs, or an alteration in the earlier cell signalling pathways, could lead to abnormal or uncontrolled cell growth and ultimately to cancer.

Two particular classes of genes have prime responsibility for controlling the lifecycle of the cells, rather as a film director and producer organize and control the activities of a crowd of film actors and extras. The so-called *proto-oncogenes* control the growth, differentiation and proliferation of the cells under their control, while *tumour suppressor* genes code for enzymes that control DNA transcription, DNA repair and other mainly nucleus-located functions. Clearly, any kind of damage to these genes, for example, by a carcinogen, virus or ionizing radiation, usually disrupts their normal functions. Under these circumstances, the proto-oncogenes become *oncogenes* and these encourage excessive growth of cells. Changes (*i.e.*, mutations) in the tumour suppressor genes disrupt their ability to destroy rogue cells. The discovery of these (proto)-oncogenes and tumour suppressor genes revolutionized research in cancer biology, and was undoubtedly a major milestone *en route* to what is hoped will be the eventual eradication of the disease.

The dawn of this new era began in 1970 with work carried out on the Rous sarcoma virus of chickens. Steve Martin and Peter Duesberg (later to become a controversial figure in the AIDS area) working at the University of California in Berkeley, and Peter Vogt at the University of Southern California, Los Angeles, demonstrated that the DNA of the Rous sarcoma virus included genetic material that was inessential for replication and reproduction of the virus. Since this virus was highly carcinogenic – if injected into chickens, it induced the growth of sarcomas within one to two weeks – the obvious presumption was that this superfluous genetic material was responsible for the carcinogenicity of the virus, that is, it had oncogenic potential, and they christened this genetic material the *src* (for sarcoma) oncogene. However, the really important discovery was made some five years later by Michael Bishop and Harold Varmus of UC, San Francisco. They showed that a DNA sequence essentially identical to that of the *src* gene was also present in normal chicken cells, and by implication the Rous sarcoma virus had 'picked up' this gene during a previous infection of chicken cells. In other words, normal cells carried genes with the potential of becoming oncogenic. In 1978, it was established that the *src* gene coded for a kinase, indicating the importance of such enzymes in the control of cellular processes.

Other oncogenes were soon discovered in a variety of animal viruses, and as the techniques of DNA sequencing became more routine, a surprising number of these oncogenes were matched up with normal genes in animals and even in humans. But what was the role of these proto-oncogenes? Most of them have now been shown to code for enzymes that help control the growth and differentiation of cells. In many instances, the enzymes are kinases that phosphorylate serine, threonine or tyrosine residues of other proteins that constitute part of a cascade of events that begins with an extracellular signal (for example, a growth factor or hormone), and ends with a major change in activity of the cell. Phosphorylation of these amino acids changes the activity of the proteins/enzymes primarily due to the associated changes in structure and charge. Clearly, any alteration to this delicately balanced programme will lead to a disruption of cellular activity.

In 1982, three research groups demonstrated, almost simultaneously, the existence of human oncogenes related to a family of proto-oncogenes called *ras*. The enzymes coded for by these genes, so-called *Ras* proteins, were all involved in translation of the primary extracellular signal into the initial secondary message just inside the cell membrane, a process controlled by the so-called G proteins (see Fig. 4.5). In the figure, a typical sequence of events is depicted. In this instance, a pair of identical growth factor molecules are shown binding to their paired receptors. This initiates the addition of phosphate groups onto the inner portion of the receptors, and this then triggers the activation of a pair of proteins called GRB2 and Sos, which in turns activates the *Ras*-G protein. This occurs through a displacement of guanosine diphosphate (GDP) and replacing it by guanosine triphosphate (GTP). Once activated, the *Ras*-G protein initiates a whole cascade of kinase-mediated reactions that ultimately send a transcription factor into the cell nucleus to interact with DNA. A burst of cell growth is then initiated. In normal cells, the *Ras*-protein is only activated (switched on) in the presence of an external signal for growth, differentiation or

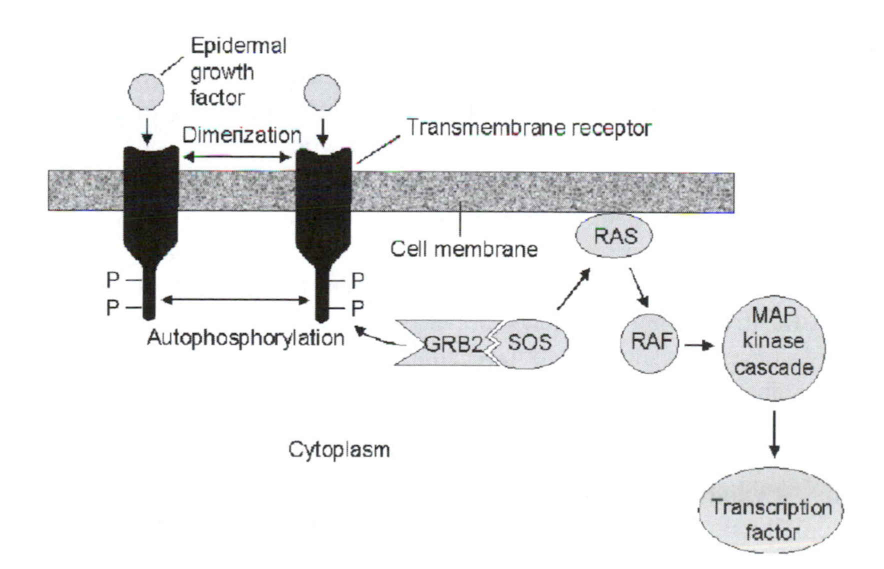

Figure 4.5 *Signalling pathways*

proliferation. The *ras*-oncogene products (that is, aberrant *ras*-proteins) could not participate correctly in this process because they were permanently switched on. What was surprising was the frequency with which *ras*-oncogenes appeared in human tumour cells: in acute myeloid leukaemia (25%), colon cancer (50%), lung cancer (50%) and pancreatic cancer (90%). In molecular terms, these oncogenes were almost universally formed by a very minor change to the DNA sequence of the proto-oncogene, which ultimately led to the production of an enzyme with one single amino acid change with respect to the native enzyme. The biological activity of this new enzyme was, however, completely different, in that it had decreased GTPase activity, thus ensuring that the GTP–RAS complex was stabilised. This ensured that the cell was permanently activated for growth and proliferation.

Many different oncogenes have now been described and most appear to ensure that cells are switched on for growth and proliferation, even in the absence of the normal stimulating signals. Oncogenes of the *myc*-family are commonly found in leukaemia cells but also in tumours of the lung, breast and stomach. These oncogenes alter the action of transcription factors in the nucleus, thus ensuring that protein production and growth are always activated. Oncongenes of the ERBB-1 and ERBB-2 family are commonly associated with breast cancer, and the parent proto-oncogenes code for receptor proteins that interact with epidermal growth factor and other growth-promoting factors.

The tumour suppressor genes normally code for the production of proteins that help to control gene transcription, DNA repair, the cell cycle, and they are also key components of biochemical pathways activated in response to DNA damage. The cell responds to such damage, by cessation of growth, by differentiation (that is, changing into a different type of cell) and by senescence (whereby the cell stops growing and dies), or it embarks upon what has become known as *programmed cell death* or *apoptosis*. This is an important process in normal foetal development whereby embryonic cells differentiate to form cells characteristic of particular tissues (bone, muscle, nerve, *etc*.). If they are incorrectly located, it is vital that they suffer programmed cell death or apoptosis. Apoptosis is clearly also important in cells that have suffered serious DNA damage. Tumour suppressor genes like the p53 gene and the retinoblastoma genes are found to be aberrant (mutated) in around half of the tumours, and this ensures that cancer cells grow in an uncontrolled way, and do not senesce, differentiate or succumb to apoptosis. It is also worth noting that the most treatable (and potentially curable) cancers, like Wilm's tumour (of the kidney), retinoblastoma (of the eye) and acute lymphoblastic leukaemia, rarely possess mutated p53 genes. Destruction of the p53 genes may also be important, and it seems that the papilloma viruses are able to degrade certain forms of these genes. Women possessing these susceptible forms of p53 may be more likely to contract cervical cancer if they are infected with the virus.

Most of these mutations by themselves will not produce a tumour and it is usually reckoned that as many as 10 such changes must accumulate before the tumour cell line is established. Even when the rogue cell population is growing unchecked, there is still no certainty that it will become a cancer rather than a localised, non-malignant growth. For this to happen, the tumour must become invasive and spread to distant sites in the body, a process known as *metastasis*. To escape from the confines of a localized tumour, the cells must evade yet more control systems and enter either the bloodstream or the lymphatic system (the fine network of tubes through which the white blood cells of the immune system are transported to all parts of the body). This is the most insidious and dangerous feature of cancer, as whereas a surgeon can excise a localized tumour and a radiotherapist can irradiate and perhaps destroy such a tumour, only chemotherapy can have any effect once the tumour has metastasized. And it is these secondary, metastatic growths in the brain, lungs, liver and bones that cause most of the pain of cancer and are the ultimate cause of death.

So how is it that normal cells stay put in the tissue where they belong (with the exception of white blood cells that are always on patrol for bacteria and damaged cells), while cancer cells can migrate? The answer seems to lie in the propensity of normal cells to release adhesion molecules, which bind them to their neighbours and to the underlying basic structure of the body – the so-called extracellular matrix. In cancer cells, these adhesion molecules

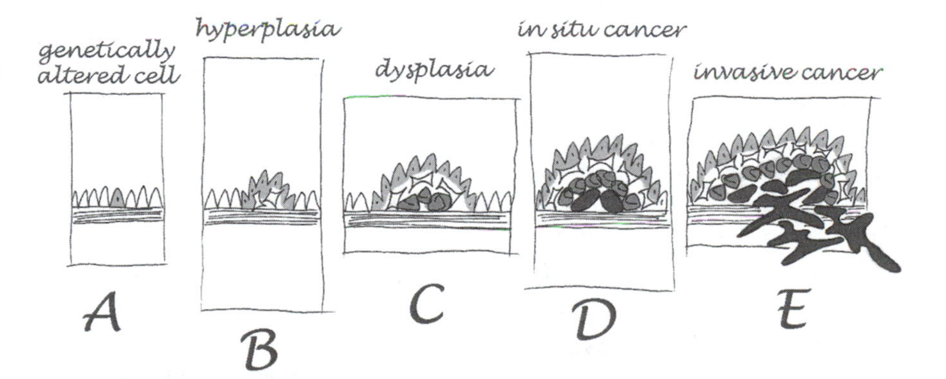

Stages in cancer development

(mainly glycoproteins) are missing or have a modified affinity for the tissues from which they arose. For example, in Burkitt's lymphoma, a relatively common tumour in Africa, the probable sequence of events is: infection with the Epstein Barr virus causes a proliferation of infected B-lymphocytes in patients who are co-infected with malaria (and thus with compromised immune systems); activation of the C-*myc* oncogene in some of these B-cells with resultant further stimulation of B-cell proliferation; and finally, loss of adhesion molecules from the surface of these proliferating B-cells, so that they evade the destructive attentions of T-lymphocytes, and the lymphoma develops.

Clearly, a greater understanding of these recognition phenomena and of all the various control mechanisms would help enormously in the design of effective anti-cancer drugs. Some of this information is beginning to emerge, but in the meantime, what steps can be taken to eradicate cancer once it is established? Despite major advances in surgery and radiotherapy, the most revolutionary changes in cancer treatment during the past 40 years have been in the area of chemotherapy.

CANCER CHEMOTHERAPY

At the time of diagnosis, a tumour will probably have a mass of one gram and contain at least 10^9 cells and the treatment that a patient subsequently receives will depend upon many factors. If the tumour mass can be reduced to about ten thousand cells with a weight of perhaps 10 μg, that is, 99.999% of all of the cells have been killed, then there is an excellent chance that the immune system will be able to assist in the total eradication of the tumour. However, even when 99.99999% of the cells have been killed, there are still 100 surviving cells, and any one of these can reproduce itself and provide the basis for a new tumour. The doubling times for tumours differ markedly; hence, for Hodgkin's lymphoma (a cancer of young people), with a doubling time of 3–4 days, the period of remission from the tumour could be very short, while for breast or lung cancer with a doubling time of perhaps 90–100 days, the period of remission could be very long.

However, herein lies the paradox of cancer treatment. Those cancers for which it is most easy to achieve a complete cure are those with the shortest doubling times. A cancer cell that is in the process of reproduction (mitosis) is at its most vulnerable and most anti-cancer drugs act at one or more of the stages of cell cycle shown in Fig. 4.4. Thus, while major advances have been made over the past 30 years in the treatment of fast-growing tumours, like childhood leukaemia, Hodgkin's disease and testicular teratoma (doubling times all in the range of 3–6 days), less success has been achieved with the major killers like breast, prostate, lung and colon cancers, with doubling times typically in the range of 80–100 days. With these slower-growing tumours, complete eradication of the very last tumour cell is much harder to achieve, and although periods of remission can last for years, it is not uncommon for the tumour to re-emerge as long as 10–15 years after remission was first achieved.

The aim of cancer treatment is thus clear: total eradication of the cancer cells must always be the ultimate goal. Major advances in surgery and radiotherapy have enormously improved the chances of attaining this goal, but once the tumour has metastasized, chemotherapy is really the only option. Progress in this area will be the subject of the following sections.

From Poison Gas to Platinum Drugs

And watch the white eyes writhing in his face,
His hanging face, like a devil's sick of sin;
If you could hear, at every jolt, the blood
Come gargling from the froth-corrupted lungs,
Obscene as cancer, bitter as the cud
Of vile, incurable sores on innocent tongues.
[*Dulce et decorum est*, Wilfred Owen, October 1917].

These poignant lines describe the effects of poison gas on a soldier caught without his gas mask. Of the various gases used during World War I, mustard gas (*bis*-β-chloroethyl sulfide) was the most insidious because the effects were less immediate than those caused by chlorine or phosgene. Mustard gas was first used at Ypres on the night of July 12th/13th, 1917, when the Germans fired 50,000 shells containing the brown liquid that smelt of garlic. The allied soldiers perceived no immediate danger from this new agent and most did not even bother to put on their gas masks. They slept in the trenches as the heavy vapour permeated their clothes, eyes and lungs. They awoke to find themselves suffering from severe eye irritation, which effectively blinded many of them, but the more insidious blistering effect of the agent gradually overwhelmed them. Initially, there was a generalized rash, which turned to blisters that would take up to four weeks to burst. If these became infected, the sores could take a further 5–6 weeks to heal. Many also suffered from throat inflammation and subsequent lung problems, leading to pneumonia and death in 1–3% of cases. By the end of July, there were 15,000 casualties, and from then until the end of the war there would be a further 125,000 casualties due to the effects of mustard gas. This figure represented 75% of all allied casualties caused by gas attacks.

Mustard gas was never intended as a killing agent but rather as a disorienting and debilitating agent. Its low volatility (unlike the other poison gases chlorine and phosgene) meant that it lingered on the battlefield and in trenches for days on end. Many senior officers considered it to be the ideal weapon of war especially for controlling errant colonials. For example, in May 1919, the War Office proposed sending mustard gas shells and other lethal agents to India as an aid to controlling the rebellious Afghan tribesmen. The India Office raised serious objections, but were over-ruled by the then Secretary of State for War, Winston Churchill, who noted in a memo that:

> Gas is a more merciful weapon than a high explosive shell, and compels an enemy to accept a decision with less loss of life than any other agency of war... . If it is fair for an Afghan to shoot down a British soldier from behind a rock and then cut him to pieces as he lies wounded on the ground, why is it not fair...to fire a shell which makes the said native sneeze.

There is no evidence that Churchill had experienced the effects of gas first hand, but he had experienced the assaults of wild tribesmen in his youth, and clearly had no love for the natives of Afghanistan.

The damage caused to the lungs, eyes and skin was generally ascribed to the production of hydrochloric acid when the agent interacted with the water in these tissues, and although there were isolated reports of toxic effects on white blood cells and in the gut, little work was done to establish the exact mode of action of mustard gas.

Between the wars, some research was carried out on war gases, but once the Second World War had begun, work began in earnest, resulting in the production of nitrogen analogues of mustard gas. Like mustard gas, this new type of agent also had vesicant (blistering) properties, but they were also toxic to many cells in the body. Most important for their subsequent use in cancer chemotherapy, this cell toxicity (*cytotoxicity*) was related to the rates at which the particular cells were dividing.

In passing, it is worth noting that a further opportunity to observe the wide-ranging effects of the agent, mustard gas, followed the bombing of the freighter *John Harvey* in Bari harbour on December 3rd, 1943. The vessel contained 100 tons of this lethal agent in the form of 100 lb bombs, but even the captain of the vessel was not supposed to know the identity of his deadly cargo. In all, 16 ships were sunk with an immediate loss of life of more than 1000, but many of the survivors were unaware of the terrible fate that awaited them as they clung to wreckage soaked in a mixture of oil and mustard gas. A subsequent Allied Command secret report noted that: "The individuals, to all intents and purposes, were dipped into a solution of mustard in oil, and then wrapped in blankets, given warm tea and allowed a prolonged period of absorption."

Once in hospital, 617 of the survivors were found to be suffering from serious eye damage, dreadful burns and severe lung complications. Some of the more seriously affected lost up to 90% of the skin and a total of 83 persons died from the effects of the agent. Some of the bodies were shipped to Porton Down in the UK or Edgewood Arsenal in the USA, for post-mortem examination. Such was the secrecy surrounding this accident, that for a while the Allied High Command believed that the Germans had carried out a mustard gas attack on the harbour at Bari.

In order to study the biology of these new nitrogen-containing mustards, the US Office of Scientific Research and Development signed a contract with Yale University in 1942. Given the nature of the work, one of the main researchers, Alfred Gilman, was a major in the US Army Medical Corps and all of them were sworn to secrecy. They quickly discovered that administration of the *N*-mustards to mice bearing tumours (mainly lymphomas) produced almost immediate effects – the tumours regressed and eventually disappeared, only to reappear about a month later. A second administration of the drugs was much less effective and the lymphomas eventually killed the animals. Such initial success followed by gradual development of drug resistance is, unfortunately, a recurring problem in cancer chemotherapy. The rapid response to these drugs was, nonetheless, unprecedented, and the Yale group was encouraged to try the drugs on human cancer patients.

In August 1942, a 48-year-old man in the terminal stages of a lymphoma was treated in the New Haven Hospital associated with Yale. He was a silversmith who had first been diagnosed with a small tumour mass in his neck in January 1941. Radiotherapy reduced the size of the mass dramatically, and he remained well until December 1941, when he received a further course of radiotherapy for a renewed growth. Rapid growth of his lymphosarcoma started in May of 1942 and there was no further response to radiotherapy. His case was considered to be hopeless, in that the tumour had failed to respond to radiotherapy and had grown to such giant proportions that it covered his chest and most of the lower part of his face. He was in constant pain and could not chew or swallow. After a 10-day course of treatment with the nitrogen mustard *tris*-(2-chloroethyl)amine the tumour mass had not only been checked but was actually regressing rapidly, and seven days later the tumour had completely disappeared. The actual case report by Gilman and Goodman and their co-workers conveys some of the drama of this study:

> "On the fourth day of treatment the patient felt better, was able to swallow and could sleep lying down. By the 10th and last day of treatment the cervical masses were no longer palpable and the axilliary masses receded completely four days later." This dramatic improvement was, unfortunately, short-lived and the tumour soon reappeared. Subsequent courses of

> treatment were less effective and the poor man died soon afterwards: "...at the time of death, three months after the start of therapy...the tumour masses were relatively small. Death was hastened by the untoward effects of the drug on the bone marrow."

This sad conclusion to the trial could not disguise the potency of the drug as an anti-tumour agent, and five other patients were subsequently treated, with similar good periods of remission.

These clinical responses were of sufficient interest to prompt the initiation of a full clinical trial of a second mustard (mustine or mechlorethamine) in more than 150 patients in a number of hospitals. Particularly good results were achieved with patients suffering from Hodgkin's lymphoma, and some of these people experienced periods of remission that lasted up to three years, provided that they had periodic doses of mechlorethamine. Less satisfactory results were obtained in patients who had other types of lymphoma or leukaemias.

It has to be said that the adverse effects of these drugs were severe, suppression of bone marrow cells being the most significant and dose-limiting (*i.e.*, life-threatening) problem. Goodman and Gilman were careful to emphasize this in their keynote report:

> The margin of safety in the use of these chemicals is narrow, necessitating the exercise of considerable caution.

Interestingly, nausea and vomiting and other general gastrointestinal problems were not prevalent, which is in stark contrast to treatment with most other types of anti-cancer drugs.

The early views on the mode of action of both mustard gas and these new nitrogen mustards were that they inactivated various cytoplasmic enzymes by causing actual chemical damage. While this kind of damage certainly occurs, an alternative suggestion was made by Gilman and his collaborator Frederick Philips in a landmark article in the April 5th, 1946 issue of *Science*. They made the observation that:

> "...threshold doses (*i.e.*, low doses) evoke pathological changes only in cells and tissues which normally exhibit relatively high rates of proliferation and growth..."; and: "...On the basis of the marked susceptibility of nuclear mechanisms (responsible for initiating cell division and growth) it is provocative to associate the cytotoxic action of the mustards with primary disturbances of nuclear function."

Given the secrecy that had surrounded these new drugs and the clinical trials during World War II, this was the first substantial report about them. Not surprisingly, it led to a flurry of research activity, not least because Gilman and Philips pointed out that: "At present only two of the nitrogen mustards have been investigated clinically...literally hundreds of congeners remain to be synthesised and evaluated."

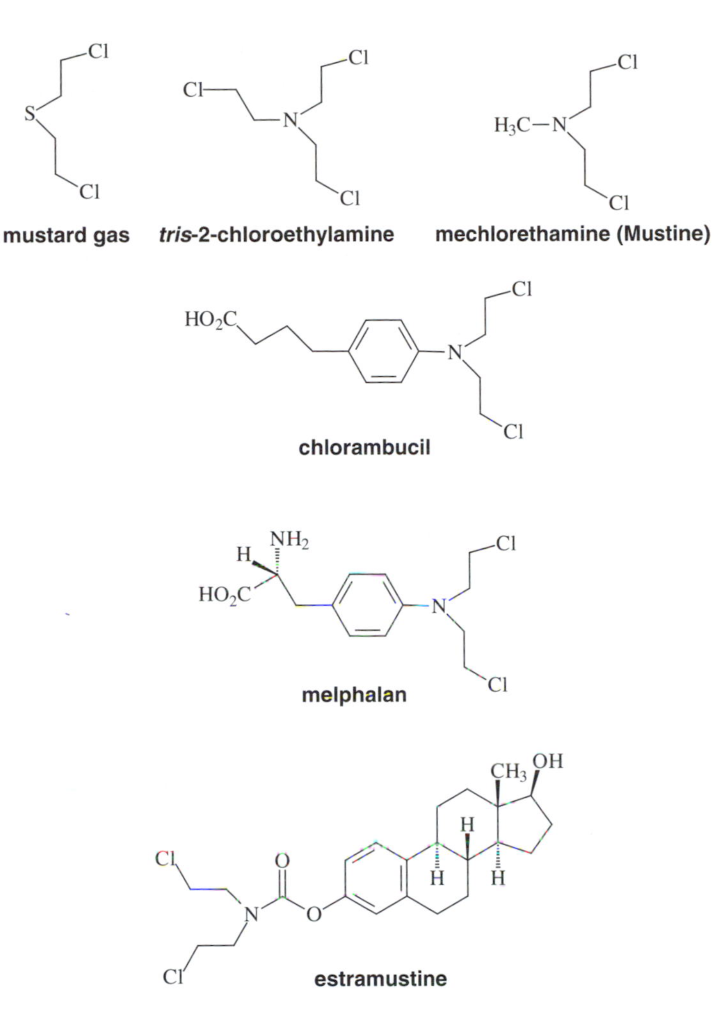

Significant improvements were achieved with the drugs chlorambucil (Leukeran, 1952) and melphalan (Alkeran, 1954), both developed at the Chester Beatty Laboratory in London. These drugs were more water-soluble (and hence available by mouth rather than intravenous infusion) and more easily taken up by cancer cells. They were also less toxic to bone-marrow cells and have been widely used over the intervening years for the treatment of various leukaemias and other tumours. An interesting combination of the female sex hormone oestradiol with mechlorethamine was invented in Romania in 1966, and was marketed under the trade-name Estracyte (estramustine) for the treatment of prostatic cancer.

But the most successful nitrogen mustard was (and still is) cyclophosphamide (Endoxana, Cytoxan, 1956) develped by the German company

Asta-Werke. This was conceived as a prodrug, in that the nitrogen mustard portion was held in the form of a cyclic phosphoramide that would need to be broken down in the body before the active mustard could be revealed. It was envisaged that this would occur through the action of enzymes known to be prevalent in prostatic tumours (acid phosphatases). In the event, the drug had excellent activity against a whole range of tumours but not against prostatic tumours. It was not, as planned, activated by phosphatases, but was activated through hydroxylation by liver enzymes, with ultimate production of acrolein and a simple nitrogen mustard (Fig. 4.6), both of which are cytotoxic.

It is interesting to compare the relative toxicities of the two extreme types of nitrogen mustard. Over 90% of mechlorethamine is destroyed within

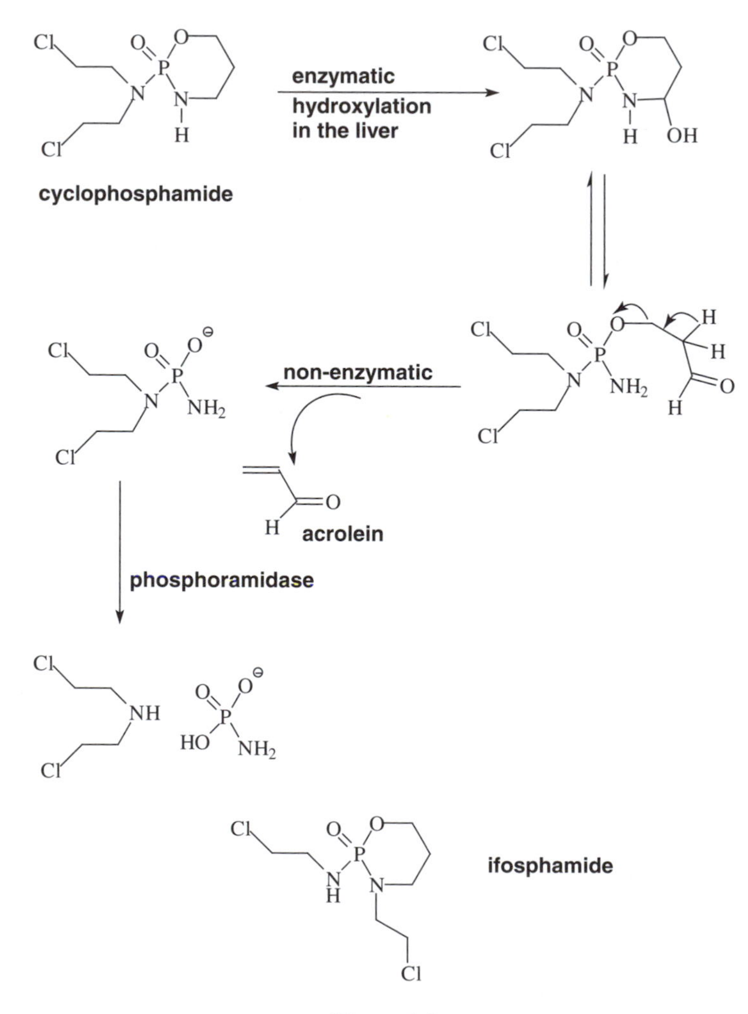

Figure 4.6

about four minutes of intravenous administration and inevitably causes much local tissue damage. In contrast, cyclophosphamide (and a similar drug called ifosfamide) are inactive and essentially non-toxic until metabolism occurs in the liver, and their anti-cancer activity only decreases slowly, dropping to half its initial value after about nine hours and to zero activity after 24–48 hours. Both cyclophosphamide and ifosfamide are administered by mouth – a major advantage for the patient – and neither of them has vesicant (blistering properties), a far cry from the actions of mustard gas. However, they do have significant adverse effects, in particular bone-marrow toxicity and most annoying for the patient, rapid alopecia. Hair loss is usually complete, and although the hair almost invariably regrows in time, there is often a loss of colour. Reduced fertility is also a problem.

One further development is worth mentioning. A slight chemical modification to the nitrogen mustards provides the nitrosourea mustards like carmustine (*bis*-chloroethyl nitrosourea, BCNU), lomustine (chloroethyl, cyclohexyl nitrosourea, CCNU), semustine and streptozocin. The first three of these penetrate the brain well and are thus useful for eradicating deposits of cancer cells there. Unlike the other nitrosoureas, streptozocin is a natural product, isolated from the microorganism *Streptomyces acromogenes*, and has selectivity for tumours of the pancreas.

All of these drugs have one thing in common – their mode of action. The double helix of DNA must unwind in order to reveal the DNA blueprint, which then acts as a template for replication of the DNA or for the formation of messenger RNA. The latter provides the code for eventual protein biosynthesis. The nitrogen mustards and related drugs react with various groups on the DNA strands, but most importantly with guanine residues (Fig. 4.7). They thus cause dramatic structural changes through the formation of inter-strand

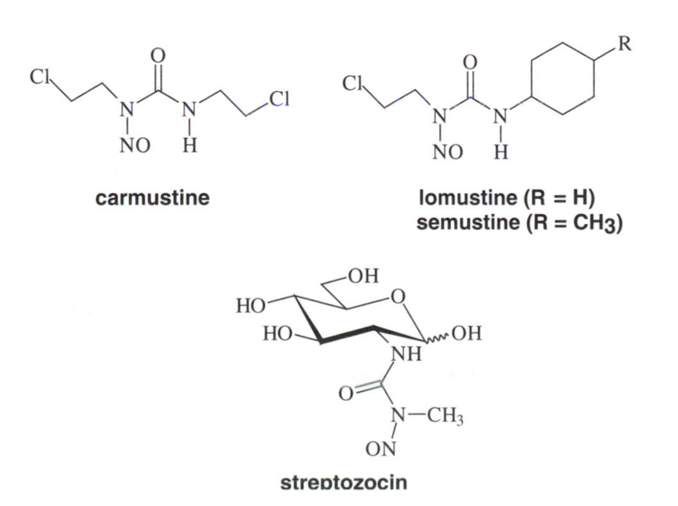

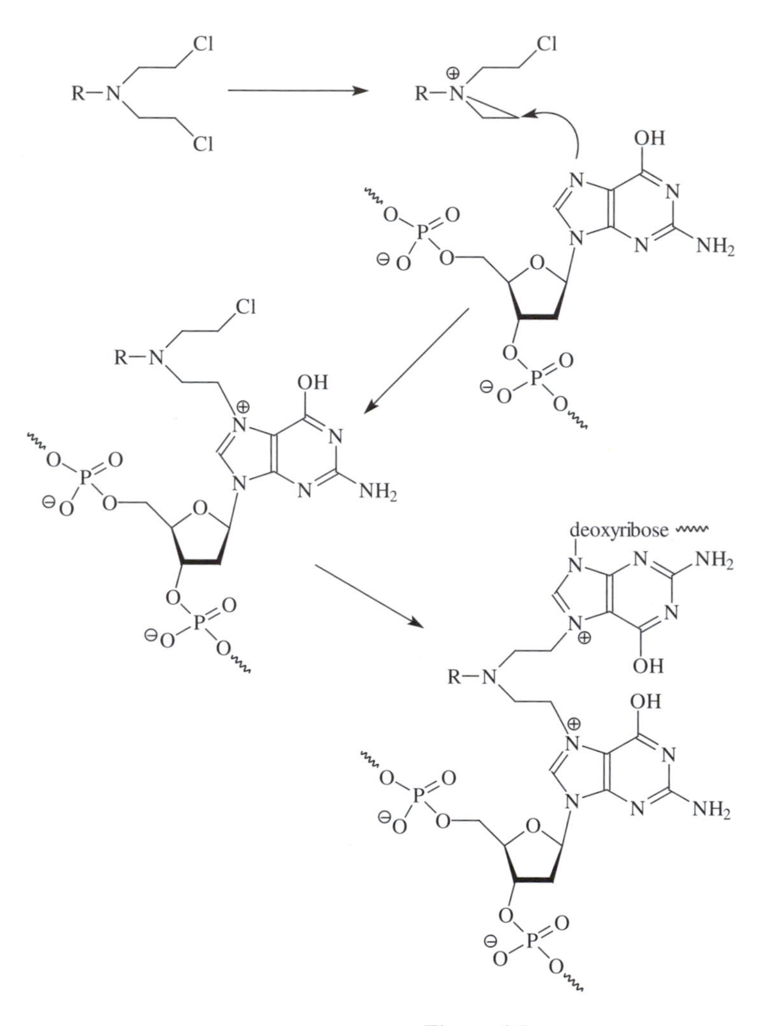

Figure 4.7

and intra-strand cross links and this prevents the helix from uncoiling and cell division and growth of cancer cells must then cease. This overall process is known as *alkylation*, and this type of drug is usually known as an *alkylator*. The unpleasant and potentially life-threatening side-effects of the drugs are probably due to their high chemical reactivity with constituents of the normal cells, like key enzymes and other proteins, and this disrupts other functions of the cells.

Both the design of nitrogen mustards and the understanding of their mechanism of action have clearly progressed a long way from those early secret studies during the war years. Paul Ehrlich would have been pleased to see his ideas of drug development arising from subtle chemical changes to existing drugs working so successfully. But serendipity can also play a large part in drug evolution, and the story of the discovery of the platinum anti-cancer agent *cis*-platin, provides an excellent example of this route to the clinic.

The background to the discovery of this drug is almost as bizarre as the discovery itself. Professor Barnett (Barny) Rosenberg was a physicist by training and in 1965 was working in the Biophysics Department of Michigan State University at East Lansing. He had recently studied photographs of the various stages of cell division taken under high magnification, and was intrigued by the similarity in appearance between the spindle fibres that held the two dividing cells together, and the patterns of the lines of magnetic force obtained when iron filings were sprinkled around the ends of a bar magnet. He decided to investigate the effect of an electric current on bacterial cells growing in culture to see if there was some kind of electromagnetic phenomenon in that situation.

His first experiment involved a pair of platinum electrodes immersed in a solution containing the common bacterium *Escherichia coli* and various nutrient chemicals, including ammonium chloride. The assumptions had been made that the platinum electrodes would be inert to both the nutrient solution and to the relatively low level of current. Fortunately, both assumptions were unjustified, and the scene was set for the discovery of an amazing phenomenon. The current was passed for about two hours, and by this time all of the rod-shaped cells had stopped dividing, but some of them had grown into filaments up to one millimetre in length, several times their normal length. Clearly, something in the solution was preventing mitosis but was allowing aberrant growth to occur.

Rosenberg and his collaborators carried out numerous experiments, changing the current, the nutrient chemicals and the pH of the solution, and ultimately concluded that the most important factors were the presence of oxygen and the ammonium chloride. The evidence seemed to point to a breakdown of the platinum electrodes and some kind of chemical reaction with the ammonium chloride under the oxidizing atmosphere provided by the oxygen. The resultant chemical species was disrupting mitosis but not growth.

Their next step was to add several commercially available platinum compounds to the bacterial cultures (but without the current or platinum electrodes), but they never observed the same disruptive pattern. Then they had another stroke of luck. One of the platinum compounds – ammonium hexachloroplatinate – proved to be unstable in solution, and on standing in the light in the presence of air, changed colour from light yellow to colourless. This solution would then reproduce the effects seen in the original experiments. A new product had evidently been produced under the influence of the light, and they proceeded to isolate this and identify it as the known compound *cis*-diammonia platinum dichloride. The pure compound reproduced the original effects at the very low concentration of about 5 ppm.

Rosenberg clearly recognised the significance of his results, and in his *Nature* paper of February 13th, 1965 posed the questions:

"Can these metal ions inhibit cell division in other bacteria, or cells?" – the inference being that they might have anti-cancer activity. Indeed, he and his senior collaborator, Loretta van Camp, tested the *cis*-platin (as it became known) against various solid tumours (especially sarcomas) in mice, and quickly showed that a single dose of about 8 mg / kg of animal injected into the peritoneal cavity (*i.e.*, into the abdominal wall) produced a complete regression of these tumours. These results were exciting, and clinical trials in patients began almost at once.

By 1973, preliminary results were available, and the good news was that *cis*-platin (soon to be known under the tradename Neoplatin) was effective against several human tumours, but was especially effective against testicular teratoma in men and a certain type of ovarian cancer (adenocarcinoma) in women. The downside was that the drug was very toxic to the kidneys and caused severe vomiting, although there was little bone-marrow toxicity like that seen with the mustards. Fortunately, the kidney toxicity could be controlled by administering large volumes of water and the diuretic mannitol along with a slow infusion of the drug over a 24-hour period, thus keeping the kidneys well flushed, and the sickness could be brought partly under control with strong anti-emetic drugs.

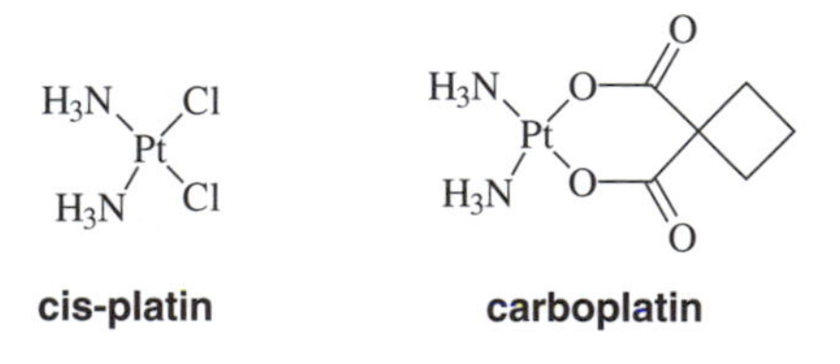

The advent of *cis*-platin improved the prognosis for teratoma patients almost beyond belief. Hitherto, patients had received surgery to remove the diseased testicle(s) and other large tumour masses, followed by radiotherapy or chemotherapy with two highly potent drugs, vinblastine and bleomycin (to be discussed later). Remissions were produced but they were often short-lived and the adverse effects of the relatively large doses of the drugs were often life-threatening (or indeed fatal). The new strategy involved surgery followed by alternating courses of Neoplatin and vinblastine/bleomycin (but in lower doses than before), until no trace of the tumour could be seen on X-ray or by estimating the blood concentrations of a protein (alpha-feto-protein) produced only by the tumour cells and by foetal cells. This latter marker proved to be a very sensitive test for the eradication of tumour cells, and clinicians were soon observing 100% response rates in their patients with two-year survival rates rising from around 20% to 60%, and with further modifications over the intervening years to the present value of greater than 95%. You now have to be rather unlucky to die from testicular teratoma*.

As for the mode of action, this is very similar to that observed with the mustards. In common with these drugs, *cis*-platin behaves as a bifunctional alkylating agent, in that it forms two covalent links to a DNA strand, causing severe structural distortion and thus preventing it from participating in replication and cell division. Most reaction takes place with the *N*-7 of guanine residues and to a lesser extent with the *N*-7 of adenine residues. X-ray crystallographic studies have proved conclusively that intra-strand bis-alkylation is more important than inter-strand cross-linking (there are probably less than 10% of the latter).

Much of the work on *cis*-platin has been carried out by the company Johnson Matthey, who have something of a monopoly on platinum supplies and technology. Over the past 20 years, a number of analogues of the original drug have been invented, and several of these are also now in clinical use. Most notable amongst these is carboplatin, which has a much lower potential for kidney toxicity and produces less nausea and vomiting. It was first licensed for clinical use in the UK in March 1986, exactly seven years after approval had been received for Neoplatin, and around 70 years after the first mustard was used as an agent of chemical warfare. *Cis*-platin remains the most useful platinum drug and is now increasingly used in combination with other drugs (like taxol – see later) for the treatment of breast and lung cancer.

BEATING LEUKAEMIA

Our blood consists of three major classes of cells: the red blood cells or erythrocytes; the white blood cells or leucocytes; and the platelets. The red blood cells contain haemoglobin and are primarily involved in the transport of oxygen around the body and in the transport of carbon dioxide and other waste products for elimination via the lungs and kidneys, respectively. The white blood cells comprise three main types: the granulocytes and monocytes that seek out, engulf and then destroy bacteria, fungi and other infecting microorganisms; and the lymphocytes, which are involved in the production of

*I have personal experience of the effects of *cis*-platin, since I was one of the first teratoma patients to be treated at the Royal Marsden Hospital with the drug in the long, hot summer of 1976. I have vivid memories of being violently and continuously sick, and then being made extremely ill by subsequent treatment with a combination of vinblastine and bleomycin. But, nearly 30 years later, I can vouch for the efficacy of the drug combination for the treatment of this type of tumour.

antibodies and the immune response. The platelets are primarily responsible for the formation of blood clots and are thus involved in wound healing.

All of these cells have their origin in a common progenitor cell (a haemopoietic cell) that is formed in the bone marrow, the jelly-like material that is at the heart of the large bones. The leukaemias are a group of cancers characterized by a disturbance in the normal differentiation and maturation of these progenitor cells, with resultant uncontrolled proliferation of a particular class of white blood cells. Perhaps the most distressing form is acute lymphoblastic leukaemia (ALL) in children, where the aberrant cells are lymphocytes. This cancer is most common in children between the ages of five and fourteen, with an incidence of about five per 100, 000 of that age group and a ratio of 8:5 with respect to boys and girls. The condition first manifests with general malaise, lethargy, severe anaemia, with boils or sores that fail to heal, and occasionally with immediately life-threatening septicaemia. Unfortunately, by the time these signs and symptoms appear, there are usually more than 10^{12} leukaemia cells. Fifty years ago, no effective treatment was available and the median life expectancy of these children was three to four months from the time of diagnosis. These days, with modern chemotherapy, virtually all leukaemic children attain a state of remission from the disease, and with subsequent use of bone-marrow transplantation and further chemotherapy, 80% or so of these children survive for five years without disease and then go on to make a full recovery. So how has this revolution been achieved?

Much of the credit for this success can be ascribed to the discovery and development of the anti-metabolite drugs. To trace the evolution of these drugs, we have to recall the excitement that surrounded the discovery of the various vitamins in the 1920s and 1930s. From the chapter on antibiotics, you will recall that the biochemistry involved in the production of tetrahydrofolate was a key to understanding the mode of action of the sulfonamides (Fig. 2.2). The fact that humans need a dietary source of folic acid from which to manufacture tetrahydrofolate, whereas bacteria must make their own tetrahydrofolate, provides the means for the selective toxicity of the sulfonamides. These successes with antibacterial chemotherapy prompted many investigations of the feasibility of using other vitamins and analogues as agents of chemotherapy.

One pioneer in this area was Richard Lewisohn who was a surgeon at Mount Sinai Hospital in New York City. He initiated an intensive programme of screening of the B vitamins in 1939. Eighty-nine mice bearing a breast tumour (an adenocarcinoma) were treated with an extract of brewers' yeast (a rich source of B vitamins and other micronutrients, including folic acid) from Germany, and 38 of these (43%) experienced complete regression of their tumours. However, another group could not replicate these results, and the outbreak of World War II meant that supplies of the yeast extract could not be obtained from Germany; hence, further research seemed impossible.

The American pharmaceutical company Lederle came to the rescue with a supply of a crystalline factor from the microorganism *Lactobacillus casei,* which they believed was folic acid. This proved to be phenomenally active as an anti-tumour agent and as little as one-quarter of a microgram caused regression of the mouse breast tumours. However, further chemical studies showed that the substance they were using was not folic acid (pteroylglutamic acid) but a derivative containing three residues of the amino acid, glutamic acid. In fact, pure folic acid proved to have no effect. Clearly, there was something special about this new triglutamate derivative, which they christened 'Teropterin'. Quantities of this drug were prepared and sent to Sidney Farber at Harvard Medical School for clinical evaluation in humans.

At this time, there was no way of knowing how folic acid and its analogues might exert their effects. Elucidation of the structure of DNA by Watson and Crick was still a decade away, and an understanding of the way in which it reproduced itself (replication) and coded for the production of proteins would not be obtained until the mid-1960s. With our present knowledge, we can state with certainty that folic acid is the precursor of dihydrofolate and tetrahydrofolate, and that tetrahydrofolate mediates the transfer of a one-carbon unit in the production of thymidine, an essential constituent of DNA. Folate analogues and antagonists thus interfere with this carbon-atom transfer and disrupt the synthesis of DNA, with a resultant cytotoxic effect on rapidly dividing (cancer) cells.

These initial studies were carried out with around 90 terminally ill patients suffering from a variety of cancers and many received up to 500 mg of Teropterin per day by intravenous administration for periods up to 35 days. Although there were no dramatic effects, some patients experienced improvement, but most interestingly, when bone-marrow biopsies were carried out on leukaemic patients, they revealed an acceleration of the disease process, and the drug probably hastened the deaths of these patients. This observation suggested two positive courses of action. One option was to administer the drug as a stimulant for the growth of leukaemic cells, which might then be rendered more susceptible to the effects of the nitrogen mustards. Alternatively, several new drugs had recently been described, as part of the sulfonamide programme, which prevented the normal functions of folic acid, and Farber and his colleagues argued that such *anti-metabolites* might have anti-leukaemic activity as well as antibacterial activity. Farber duly experimented with a number of these folic acid antagonists in 1947, but had no major success until he tried the drug aminopterin supplied by the American Cyanamid Company. Sixteen children, seriously ill with acute lymphoblastic leukaemia, received the drug during the six-month period starting in November 1947. Ten of these went on to achieve complete remission from the disease with white cell counts returning to normal. The enormous significance of this success is hard to appreciate now, since complete remission in ALL is the norm with modern chemotherapy; but in 1948, the result caused great excitement. Farber maintained his patients on low (maintenance) dosages of aminopterin and many of them survived for eight months or more, with perhaps one in a 100 going on to make a full recovery. Clearly, the drug was reducing the number of leukaemic cells to an undectable level (now estimated to have been about one billion cells), and for most patients, these reproduced and eventually became resistant to the drug. But for the rare, lucky patients, their immune system was able to destroy the billion cells before they had a chance to multiply. Inevitably, the drug produced adverse effects, most notably ulceration of the mouth due to inhibition of the cell division of oral cavity cells, but the patients also ran a serious risk of infection, since their immune sytems were seriously compromised. A safer drug, methotrexate (also an antagonist of folic acid), soon replaced aminopterin, and is still one of the best folate antagonists.

The success achieved with folic acid antagonists encouraged others to investigate the potential clinical utility of alternative anti-metabolites. Of these, the purines and pyrimidines seemed ripe for examination. It was known that some of these were key components of the structure of DNA, and we have already seen how disruption of the replication or production of DNA (by mustards and platinum drugs) can have a serious effect on cancer cells. Various structural analogues of the essential purines and pyrimidines had already been screened as anti-metabolites, the theory being that incorporation of the incorrect (*i.e.*, analogue) forms of these species into growing DNA strands would produce aberrant and hence non-functional forms of DNA. Of the numerous analogues tried, 6-mercaptopurine (first tried in 1952) proved to be the most efficacious for the treatment of ALL, with periods of remission lasting for a year or more. This was one of many drugs prepared by the Wellcome Research group of Gertrude Elion and George Hitchings, whose other discoveries included the immunosuppressant azathioprine and the best-selling anti-viral agent acyclovir. For this seminal research, carried out over a period of three decades, they shared the Nobel prize for Physiology or Medicine in 1988 (together with James Black for his work on the beta-blockers and H2-receptor antagonists).

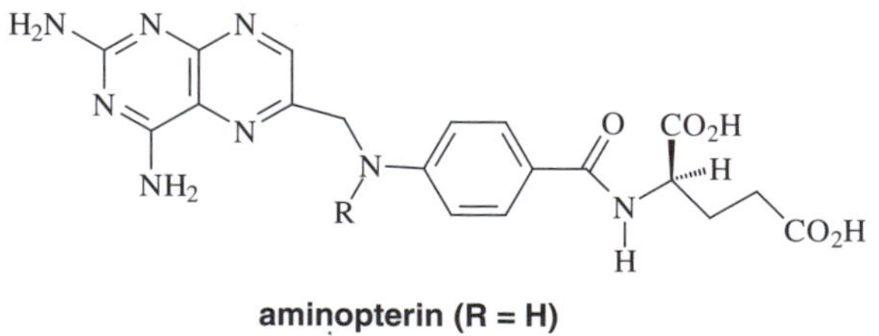

aminopterin (R = H)
methotrexate (R = CH_3)

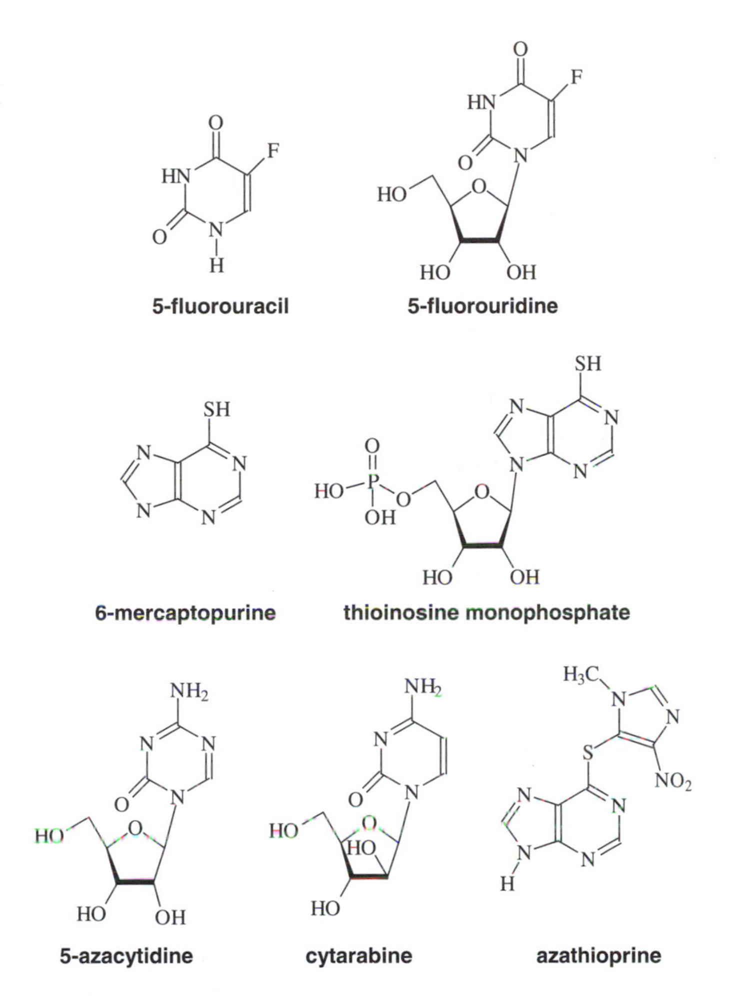

It has subsequently been shown that mercaptopurine is first metabolised by the enzyme hypoxanthine phosphoribosyl transferase into thioinosine monophosphate and this then acts as an inhibitor of *de novo* purine biosynthesis. Mercaptopurine worked even in patients whose leukaemia cells had become resistant to the effects of methotrexate, suggesting that combination chemotherapy with the two drugs might extend the period of remission from disease. The idea of combination chemotherapy to delay induction of resistant strains of microorganisms had, it will be recalled, been suggested by Ehrlich in around 1912. With leukaemia, it was the major breakthrough that was needed to achieve long-term survival in the disease.

During the 1960s and 1970s, various medical centres throughout the world began to experiment with a number of aggressive forms of combination chemotherapy. Their patients were given various mixtures of drugs, of which methotrexate, mercaptopurine, vincristine (a natural product from the plant

Catharanthus roseus, which is a potent inhibitor of cell division, and will be discussed in the next section) and prednisone (a steroid) were popular and highly effective examples. This combination induced complete remission in up to 95% of patients. One deficiency of this chemotherapeutic regimen was the failure of these drugs to penetrate the brain, and this would provide a reservoir of leukaemia cells that could reproduce, repopulate the body and kill the patient. The American physician Donald Pinkel, of St. Jude's Children's Hospital in Memphis Tennessee, neatly solved this problem by irradiating the skulls of his patients with X-rays (*i.e.*, radiotherapy) in order to kill this hidden population of cells.

Thirty years on, this basic approach has not changed dramatically. Three stages can be distinguished in modern chemotherapy for ALL. The first stage involves induction of remission using a combination of vincristine, prednisolone, methotrexate and mercaptopurine. There follows a period of (prophylactic) radiotherapy of the brain to ensure complete eradication of all leukaemic cells, and then an extended period of maintenance therapy using methotrexate and mercaptopurine. Around 95% of the patients achieve remission using this drug regime. More recently, a more effective combination has been used. This comprises the natural product doxorubicin (from the mould *Streptomyces peucetius* – to be discussed later in the chapter), which interferes with DNA replication, and asparaginase, an enzyme that destroys the amino acid asparagine, which is required for the growth of leukaemia cells. This type of enzyme activity was first identified in guinea-pig serum by Clementi in 1922; but the anti-tumour activity of this serum was not demonstrated until 1953 by John Kidd working at the University of Cornell. He was impressed with this "naturally occurring substance that brings about regression of cancer cells in living animals without obvious harm to the latter." A colleague, John Broome, noted the obvious link with Clementi's work, and purified the asparaginase for further clinical evaluation. Guinea-pig serum was not a convenient source for large-scale isolation, but in 1964, the bacterium *Escherichia coli* was shown to be a much more useful source of the enzyme. Although the anti-tumour potential of asparaginase has been assessed in a wide variety of tumours, it has never shown useful activity except in leukaemia.

The optimum period for maintenance therapy is usually believed to be 30–36 months. Once all treatment has ceased, the patient can be considered to be cured if the period of complete remission is maintained for four years or more. Eighty-five percent of patients can expect to attain this status – a far – cry from the virtually zero chances of survival 50 years ago.

Other Anti-metabolites

Over the years, hundreds of purines and pyrimidines have been prepared and tested for their anti-tumour activity (and more recently antiviral activity). Of these, 5-fluorouracil (5-FU) and cytarabine have been particularly useful. The former was first synthesised by a research group at Hoffmann LaRoche, Nutley, New Jersey, and then tested on transplanted tumours in mice and rats by Charles Heidelberger of the McArdle Memorial Laboratory at the University of Wisconsin in Madison, in 1957, and experience over the intervening years has shown that it has particularly good clinical effects in cancers of the colon and rectum. It is interesting to recall why Heidelberger chose 5-fluorouracil. He believed that uracil (or a biological derivative) was the precursor of thymine and/or thymidine, a key component of DNA. He further reasoned that since replacement of a hydrogen atom in acetic acid by a fluorine atom produced a new molecule (fluoroacetic acid) that was highly poisonous, through a disruption of acetic acid metabolism, the same change in uracil might have a similar effect. With our present understanding of DNA production, it is clear that the Heidelberger's ideas were highly perceptive. The actual mode of action involves inhibition of the enzyme thymidylate synthetase as shown in Fig. 4.8. The enzyme normally reacts with deoxyuridine monophosphate (X=H in the FIG) in a Michael reaction and the resultant enolate then attacks methylenetetrahydrofolate. The key step then involves loss of a proton ($X=H^+$), followed by the release of tetrahydrofolate and reduction to yield deoxythymidine monophosphate. With the drug 5-FU, this is first metabolised to 5-fluorodeoxyuridine monophosphate, and after reaction with thymidylate synthetase, the mechanism requires loss of F^+ and since this is electronically impossible, the enzyme becomes irreversibly bound to the 5-FU-methylene tetrahydrofolate adduct.

Cytarabine was first prepared in 1959 and was modelled on some natural compounds that had been isolated from a species of Caribbean sponge, *Cryptotethya crypta*, in 1951. The interesting structural feature of these so-called 'spongonucleosides' was that they resembled the natural building blocks (nucleosides) of DNA, except that they contained the unusual sugar arabinose rather than the sugar deoxyribose of DNA. It was of obvious interest to test these and similar (synthetic) compounds as anti-metabolites. Cytarabine was soon shown to have the best spectrum of anti-tumour activity and was later shown to have good activity in drug combinations used to induce remission in ALL.

An interesting spin-off from these studies of anti-metabolites was the discovery (by the Elion and Hitching's group) of the immunosuppresant drug azathioprine (Immuran) in the mid-1950s. This was one of several compounds prepared as analogues of mercaptopurine that would have greater stability in the bloodstream. Its activity *in vitro* was very good, but it had little activity as an anti-tumour agent in clinical trials. It did, however, suppress the immune response to foreign cells, and Roy Calne, a British surgeon working at Harvard Medical School pioneered its use in human transplantation surgery. Organ rejection is always the greatest problem in this type of surgery, and even when the tissue match is very good, there is always enough discrepancy

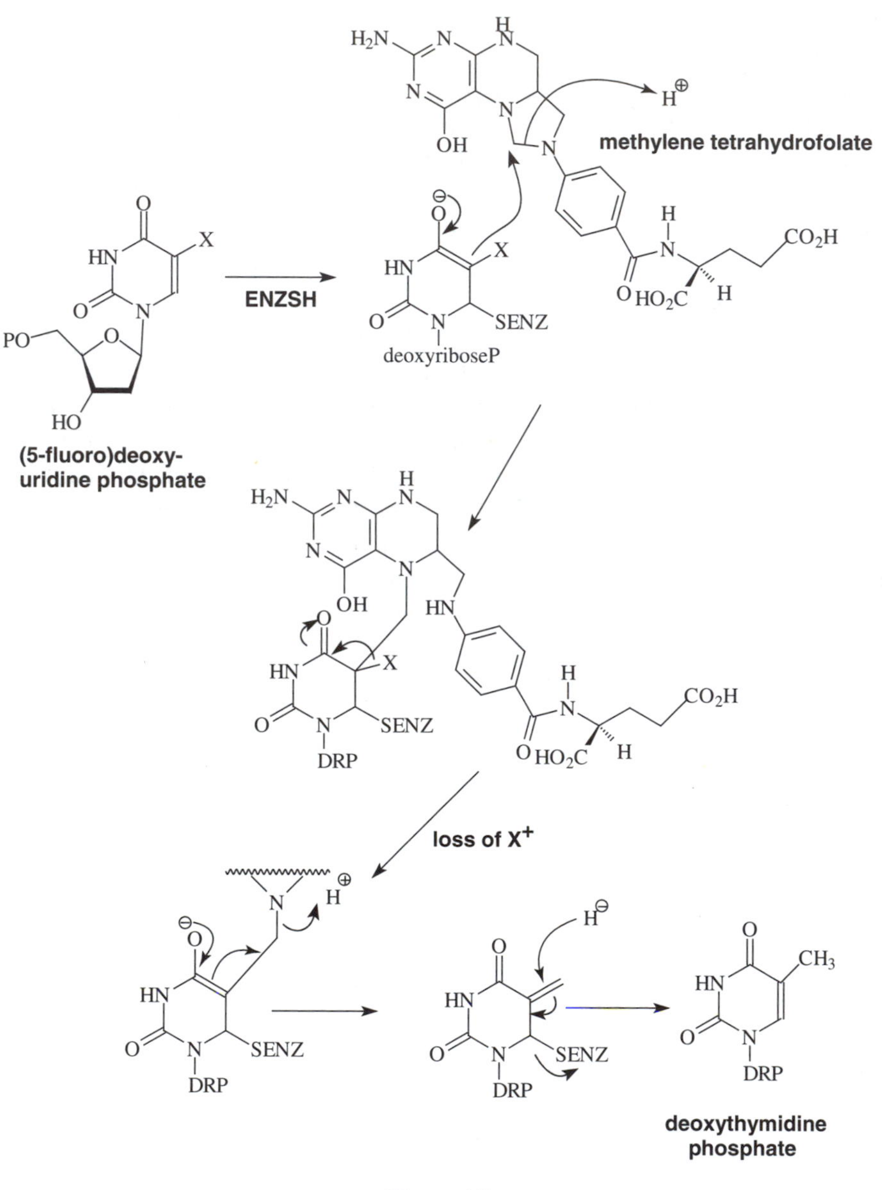

Figure 4.8

to cause an immune response in the recipient. Azathioprine has now been largely superceded (except for kidney transplants) by the drug cyclosporin, first isolated in 1972 from the mould *Tolypocladium inflatum* (see the next chapter). Again, use of the new agent was pioneered by Roy Calne, this time working at Addenbrooke's Hospital, Cambridge. It is now routinely given to patients receiving organ transplants and to those (especially children) receiving bone marrow grafts as part of leukaemia therapy. The strategy here is to administer to a patient (already in remission) whole-body irradiation with X-rays at a very high dose in conjunction with the cytotoxic drug cyclophosphamide. This near-lethal combination is designed to kill any residual leukaemia cells. The patient then receives a bone-marrow donation from a brother or sister with a near-identical tissue type. These non-diseased cells now multiply, repopulate the patient's marrow and thus provide a source of blood cells of all types. In addition to cyclosporin (for suppression of the immune response to these foreign cells), antibiotics must also be administered to counteract the possibility of infection in the immune-compromised patient. Typically, there is a 50% success rate for these procedures.

While emphasis has been given to the use of anti-metabolites in the treatment of acute lymphoblastic leukaemia in children, they are also used with varying success with other types of cancer. For the other common forms of leukaemias – acute myeloid and monocytic leukaemias – which affect adults in the age range 55–75, a cocktail of drugs that includes cytarabine, 6-thioguanine (an analogue of 6-mercaptopurine) and daunorubicin (a structural relative of doxorubicin, discussed later) can induce remission in up to 80% of patients. Bone-marrow transplantation has also become a standard part of the treatment, once remission has been achieved, and the resultant rate of cure is usually around 50%, especially in patients under 65 years.

In addition, 5-FU and methotrexate are used to treat certain forms of breast cancer, and methotrexate is the drug of choice for the treatment of choriocarcinoma in women. It is also effective in the treatment of psoriasis, a disease in which there is an over-proliferation of epidermal cells in the skin. Finally, it is worth recalling that azidothymidine (AZT) was first synthesised in 1964 as a potential anti-metabolite for cancer chemotherapy, but proved to be ineffective in this role. Its subsequent efficacy as an inhibitor of viral reverse transcriptase in the treatment of HIV infections has assured its place in the history of the 20th century.

None of these drugs can claim to be a 'magic bullet' since there are serious, dose-limiting adverse effects associated with their administration. However, like the alkylating drugs discussed in the previous section, there is often some selective toxicity towards cancer cells, and there are countless thousands of children (in particular) who owe their lives to these drugs.

PERIWINKLES, MAYAPPLES AND YEW TREES: ANTI-CANCER DRUGS FROM PLANTS

Human cultures have always experimented with plant extracts in their search for food and medicines. The Ancient Egyptians left a record of their herbal knowledge in the form of a 1 by 68 foot scroll, discovered by Georg Ebers in 1862 (now known as the Ebers' papyrus) that dates from about 1500 BC. This lists around 800 remedies, many based upon the use of plant extracts. Another ancient text (dating from about 200 BC) was the Chinese Shennung herbal, which described 365 drugs. Many well-known plants were represented in this list, including the opium poppy, the castor oil plant and *Ephedra sinaica* (which they used for bronchial problems). We now know that this contains the bronchodilator and decongestant drug ephedrine.

Knowledge of the beneficial effects of plant (and animal) extracts was passed down the generations, but it was not until the 19th century that the chemical constituents responsible for the useful pharmacological properties could be isolated in pure form. The chemical structures of most of these (*e.g.*, morphine, quinine) were so complex that they were not elucidated until well into the 20th century, although a few of the simpler ones, like atropine, cocaine and ephedrine, were solved much earlier. A systematic search for biologically active natural products from plants (and microorganisms) did not begin until the late 1940s. However, these efforts were almost immediately rewarded by the discovery of a wealth of antibacterial substances and several highly potent anti-tumour agents. Of this second class, two have been of almost unparalleled value – vinblastine and vincristine.

As so often, serendipity played a large part in the discovery process. For many centuries, the people of the Philippines had used extracts of the tropical plant *Catharanthus roseus* (formerly *Vinca rosea*, also called the Madagascar periwinkle, although the plant is not of the periwinkle family) for the treatment of diabetes. They believed that these extracts produced an alleviation of the symptoms of the condition, presumably by reducing the concentration of glucose in the blood. This was not an isolated piece of folklore, since extracts of the plant were used for the same purpose in South Africa and were marketed under the tradename 'Covinca'. In England, a preparation was sold under the tradename 'Vin-q-lin'. These various uses attracted the attention of the endocrinologist Ralph Noble working at the University of Western Ontario in Canada. Supplies of the plant were readily available in the USA and Canada, because the species had been popular as an ornamental plant for some years. Noble administered crude extracts of *Catharanthus roseus* to rats (initially by mouth and later by injection) and then measured their blood sugar concentrations. He observed no significant changes, although most of the rats died from infection when the extracts were injected. Reasoning that this implied a lowering of their white cell counts, he measured them and sure enough, they were much reduced. This was an effect similar to that seen with the new anti-leukaemic drugs, and it thus became of great interest to purify the extracts and to identify the chemical species that produced the effect. With the assistance of a British chemist, Charles Beer, Noble managed to obtain (in 1958) one pure compound that they called vincaleukoblastine – later vinblastine. This did indeed produce severe depletion of white blood cells (leukopenia).

They reported the results of their research at a meeting in New York in 1958 and were approached by a group of scientists from the pharmaceutical company Eli Lilly who had made similar discoveries. This group, under the direction of Gordon Svoboda and Irving Johnson, had also screened *Catharanthus roseus* for anti-diabetic effects, also with negative results, but had then shown that plant extracts extended the lives of mice carrying the leukaemia cell line known as P1534. It is worth noting that this particular tumour cell line was exquisitely sensitive to the plant extract, and but for this fortuitous choice, and the resultant excitement over the observed effects, the rest of the research might not have been conducted. They went on to isolate three other anti-tumour substances: vinleurosine (in 1958), vincristine and vinrosidine (in 1961). A further 70 so-called vinca alkaloids were isolated and investigated during the next 10 years. These compounds were screened against some 20 tumour lines, but only vinblastine and vincristine had really potent activity.

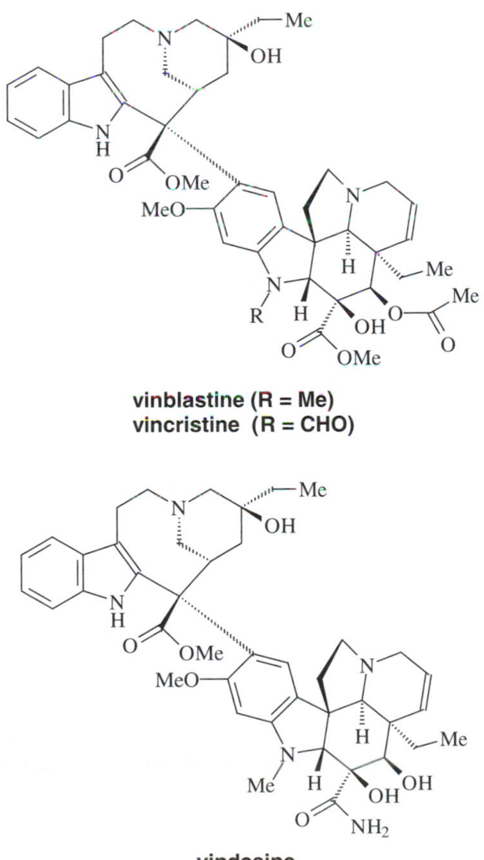

vinblastine (R = Me)
vincristine (R = CHO)

vindesine

The first reports of clinical trials in patients began to appear in the early 1960s, and these broad-spectrum anti-cancer agents soon became popular and effective drugs for the treatment of a wide range of cancers. Summarizing the experience gained by the Lilly group in 1963, Johnson wrote:

> In spite of their close (chemical) similarity, a somewhat different group of human neoplasms responds to these compounds, and there has been a singular lack of cross-resistance between these two drugs and other oncolytic drugs now in wide use.

The use of vincristine in the combination chemotherapy of both childhood and adult leukaemias has already been mentioned, and its other major success has been as part of the curative treatment of Hodgkin's lymphoma. This is a condition of (mainly) young people (with a peak in the age range 15–25) in whom there is an uncontrolled proliferation of the white blood cells known as lymphocytes with concomitant enlargement of lymph nodes. From about 1966 onwards, various combinations of drugs, all containing vincristine, were evaluated and from these studies, the MOPP regimen emerged as the most likely to induce remission and produce long-term survival and cures. This comprises a nitrogen mustard, Oncovin (the tradename of vincristine), procarbazine (a drug that causes breaks in the DNA strands) and the steroid prednisone. Vincristine is also used with great success in the treatment of a number of rare childhood cancers: Wilms' tumour (of the kidneys), neuroblastoma (a brain tumour), rhabdomycosarcoma (of muscle) and Ewing's sarcoma (of the bones). Vinblastine is also used with great effect for the treatment of Hodgkin's lymphoma, in combination with daunomycin, bleomycin (both discussed in the next section) and dacarbazine (a drug that methylates DNA, thus changing its overall structure and properties). However, the major success for vinblastine, beginning in the 1970s, was its use in combination with *cis*-platin and bleomycin (which induces the production of oxygen radicals and consequent destruction of DNA) in the curative treatment of testicular teratoma. It is also used in curative therapy of bladder cancer, in combination with methotrexate, daunomycin and cyclophosphamide.

As usual, there are serious adverse effects associated with the use of these drugs. Vincristine is the most toxic and causes a range of neurological problems (neuropathy) producing paraesthesia, that is, a loss of sensation in the fingers and toes, and problems in the gut with associated constipation and severe abdominal pain. With vinblastine, this neuropathy is less serious, and the main dose-limiting problem is depression of the number of bone-marrow cells with associated anaemia and susceptibility to infection. One structural analogue of vinblastine, vindesine, has also proved to be an effective drug (rather similar in its spectrum of activity) and with slightly less associated neuropathy and bone-marrow suppression. But how do they work?

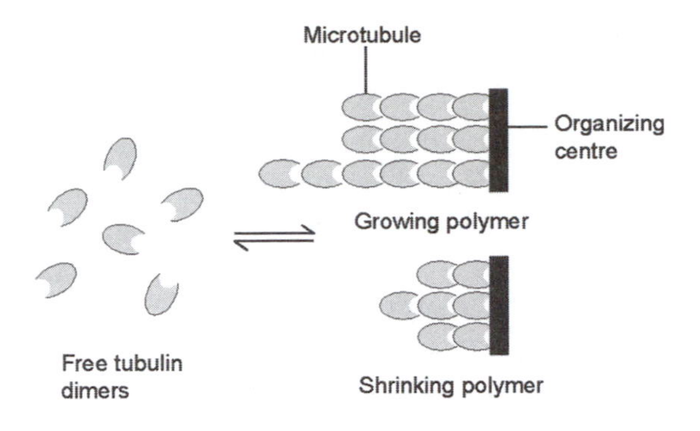

Figure 4.9 *Microtubule polymerization/depolymerization*

During cell divison (mitosis), the two developing daughter cells are held together by thin fibres called *microtubules*. These are polymers made from a protein known as *tubulin*, which exists in two similar though non-identical forms, the so-called alpha- and beta-subunits. These associate to form heteroduplexes, which then join together in a 'head-to-tail' fashion as shown in Fig. 4.9. The resultant protofilaments then polymerize to produce the microtubules. At any time, there is an equilibrium between microtubules and tubulin, assembly and disassembly being finely balanced. In the presence of vinblastine and vincristine, existing microtubules break down to produce discrete tubulin subunit pairs, which bind to the drug and are then not available for the production of new polymer. Eventually, the tubulin–drug complexes associate to form giant pseudo-crystalline aggregates, and microtubule formation ceases altogether. Cell division comes to a halt and the immature daughter cells die. Hence, cell proliferation ceases.

This mode of action is not unique since the natural product colchicine, from the autumn crocus *Colchicum autumnale* (once used as a treatment for gout), also binds to tubulin and disrupts microtubule assembly. It was never seriously considered as an anti-cancer drug owing to its low *therapeutic index*: that is, the dose required to produce clinical benefit was not much greater than the dose causing serious adverse effects or even death. A third plant product with similar biological properties, podophyllotoxin from the American Mayapple, provided the stimulus for research that led to the discovery of another excellent anti-cancer drug – etoposide.

The *Leech Book* of Bald (ca. AD 950), a herbalist who lived in the time of Alfred the Great, contains a wealth of plant lore that includes ointments to protect against "the elfin race and nocturnal visitors" , but also mentions an extract of the wild chervil (probably *Myrrhis odorata)* as a salve for the treatment of tumours. This plant produces a number of chemicals known as lignans, which are related in structure to podophyllotoxin, although much better sources of these cytotoxic lignans are the Himalayan plant *Podophyllum*

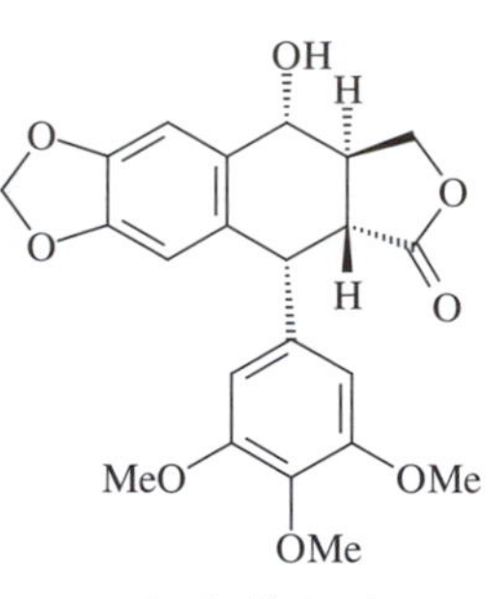

podophyllotoxin

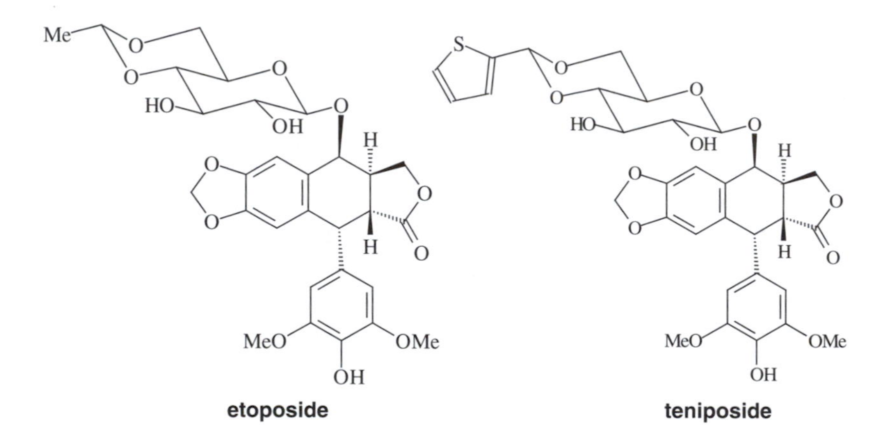

etoposide **teniposide**

emodii and the American Mayapple *Podophyllum peltatum*. The Penobscot Indians of Maine in the USA are said to have used an extract of the root of this plant as a treatment for venereal warts, and indeed, tincture of podophyllum is still available as a proprietary treatment for skin warts and verrucas.

In 1942, Kaplan, a surgeon working in New Orleans, successfully treated a venereal wart with a topical application of an extract of *P. peltatum* and this led to trials on various cancer cell lines in animals. Much of this work was carried out by the group of Jonathan Hartwell at the National Cancer Institute in Washington, DC, as part of what was to become a vast screening programme for anti-cancer agents. During the period from 1955 to 1980, the Cancer Chemotherapy National Service Center (in Bethesda near Washington) screened around 500,000 plant and microbial extracts together with wholly synthetic compounds for their anti-tumour activities.

A limited clinical trial of podophyllotoxin and a co-occurring lignan, α-peltatin, was carried out in the early 1950s, but although some short-lived responses were observed, the general toxicity of the two compounds precluded further studies. At about the same time (1954), the Swiss pharmaceutical company Sandoz embarked on a programme of chemistry directed towards the preparation of a range of structural analogues of these natural lignans. In particular, they were interested in the fact that the plant contained other lignans that had sugar molecules attached to their basic structures. These glycosides were more water-soluble than the parent lignans, usually a useful attribute in a drug, and they were shown to inhibit mitosis in cell cultures as well as in mice bearing various tumours. In addition, they appeared to cause fewer adverse effects, at least in these animal tests, although they proved to be less useful in a clinical trial. Undeterred, the Sandoz researchers made numerous other structural analogues and finally they were rewarded by the discovery of two drugs with outstanding profiles of activities in animal tumours. These were VM-26 (later teniposide), first prepared in 1965, and VP16-213 (later etoposide). Clinical trials began in Europe in 1972 and it quickly became apparent that these two drugs had broad-spectrum activity against a wide variety of tumours. They have been used as part of combination chemotherapy for the treatment of acute non-lymphoblastic leukaemia, Hodgkin's disease, bladder cancer and certain brain tumours. But the main current use for etoposide is as a less toxic replacement for vinblastine in the treatment of testicular teratoma, and as one of the few drugs valuable for the treatment of small cell lung cancer. Teniposide has a special value for the treatment of acute lymphoblastic leukaemia and in the rare childhood cancer neuroblastoma.

While the ability of podophyllotoxin to inhibit cell division was established in the mid-1940s, the mechanism of action of etoposide and teniposide was not elucidated until the mid-1980s. Like the vinca alkaloids, podophyllotoxin leads to a breakdown of microtubules and prevents polymerization to produce new ones. Etoposide and teniposide have no affect on microtubule assembly, even at doses 20 times higher than the amount of podophyllotoxin required to achieve this. They do, however, cause breaks in DNA strands, and this seems to be due to inhibition of an enzyme called topoisomerase II. In order to understand the normal functions of this enzyme, we have to consider the way in which DNA is packed into the cell nucleus. If all of the DNA of our chromosomes (we have 46 chromosomes in the nucleus of almost all of our cells) was laid end to end, it would stretch to two metres in length. Normally, the two DNA strands are wound around one another in a double helical configuration, and this is further super-coiled to provided a very compact strcuture. However, in order to reproduce itself, the DNA must become partially unravelled, and the main role of the enzyme topoisomerase II is to cause breaks in the DNA strands to allow a degree of reorganization to occur.

The type II topoisomerases catalyse the cleavage of the complementary strands of one DNA duplex, the subsequent passage of a second duplex through the opening and finally, resealing of the cleaved strands. This has the effect of changing the overall topology of the supercoiled DNA, allowing the

From and Imperial Cancer Research Fund leaflet encouraging self-examination

formation of a more open structure, which is a requisite for replication and translation. Etoposide and teniposide appear to interact with the enzyme–DNA complex and stabilize it, thus preventing topological change and eventual rejoining of the broken strands. The drugs consequently lead to a disruption of replication and cell division just as effectively as with the vinca alkaloids. They are not quite the 'magic bullets' envisaged in the Leech books (as deterrents for elves and goblins), but they are very effective anti-cancer agents nonetheless.

A number of other plant products have useful clinical activity and camptothecin and taxol are two of the most successful. The former arose out of a screening programme that took place at the US Department of Agriculture Laboratory in Philadelphia under the direction of Monroe Wall. During the period 1950–1959, thousands of plant extracts were screened for a special type of steroid that could act as a precursor of the anti-inflammatory steroid cortisone. Some testing for antibiotic and anti-tumour activity was also carried out and in 1957; about 1000 samples were sent to Jonathan Hartwell at the Cancer Chemotherapy National Service Center in Washington, since he had just embarked on an ambitious screening programme for anti-tumour agents. Between 1960 and 1987, 3500 plant samples were screened against the mouse leukaemia cell lines L1210 and P388. One of these extracts, from the Chinese tree *Camptotheca acuminata*, was shown to have very potent activity against the mouse leukaemia L1210 – the lives of the test animals were extended by up to one-and-a-half times at a dose level of around 0.5 milligrams per kilogram of mouse. The chemical structure was determined in 1966 and clinical studies began soon afterwards. These early investigations were rather disappointing probably because camptothecin underwent chemical changes during formulation, and as a result, lost much of its anti-tumour activity. Clinicians lost interest in the drug until about 1985, when its mode of action was elucidated. Camptothecin interferes with the functions of the enzyme topoisomerase I, an enzyme that has some similarities to topoisomerase II, but produces topological change by cleavage of single DNA strands and passage of other single DNA strands through the breach. During the past 10 years, new, more sophisticated trials have been conducted, and camptothecin and several of its structural analogues have proved to be quite effective against several types of cancer. It is of particular value, in conjunction with *cis*-platin, in the treatment of non-small-cell lung cancer. Interestingly, it has recently been shown to have potent activity against the parasites that cause sleeping sickness and malaria. A remarkable resurrection of interest in a drug that has been under evaluation for 40 years, and two of its analogues – topotecan (Hycamptin) and irinotecan (Camptosar) – have been approved for clinical use and exhibit much less bladder toxicity than camptothecin itself.

Probably the most successful drug to emerge from the screening programme by (what became) the National Cancer Institute in Washington, DC, was taxol from the Pacific yew tree. The yew tree has always been associated with death. The Druids planted it in their burial grounds; the ancient Greeks and Romans used it as a source of poison; and its wood provided the structure of that unrivalled mediaeval 'killing machine', the English longbow. But now it is associated with the prolongation of life. The leaves of the yew are poisonous to most mammals and a lethal dose for humans is said to be about 4–5 handfuls of the needles. This toxicity is primarily due to the presence of a group of alkaloids known collectively as taxanes. However, it was the presence of another natural product in the bark of the Pacific yew (*Taxus brevifolia*) that excited the interest of the National Cancer Institute screening programme. In 1964, a crude extract of bark was discovered to be toxic to a mouse leukaemia cell line, and purification and structure elucidation of the compound responsible were undertaken. After considerable efforts by the group of Monroe Wall and Mansukh Wani of the Research Triangle Institute in North Carolina, taxol was identified as the active constituent in 1971.

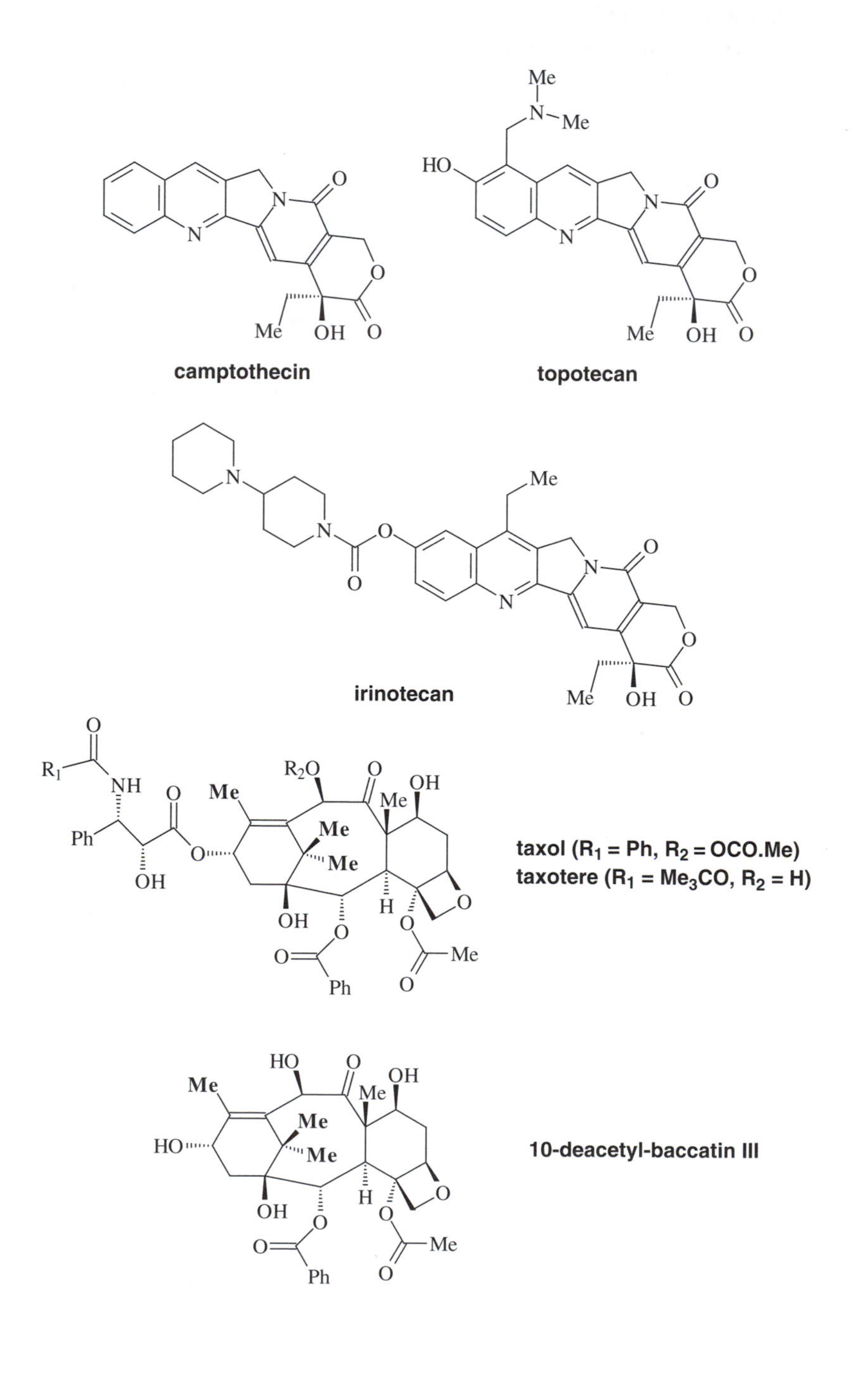

Despite the interesting pharmacology of taxol, this group did not actively investigate the anti-cancer potential of the compound primarily because it was only available in small quantities from the tree bark (typically 12 Kg of bark provided 500 mg of taxol). The Pacific yew was primarily confined to the Western states of the USA, and of limited occurrence. It was also slow-growing and the process of bark-stripping risked killing the tree. The commercial and environmental difficulties seemed insurmountable. Taxol would have remained just another natural product with an intriguing chemical structure but for the studies of Susan Horwitz of the Albert Einstein College of Medicine in New York City. She reported in 1979 that taxol disrupted cell division, although not by interfering with microtubule assembly like the vinca alkaloids, but by stabilizing them so that they could not depolymerize. Since the control of microtubule assembly and disassembly is so vital at all stages of cell division, this novel mode of disruption caused great excitement. Further biological testing also demonstrated that taxol had a better profile of anti-cancer action against human tumour cell lines (transplanted into mice, so-called xenografts) in comparison with the activity against mouse tumour cells. It was thus of considerable interest to take taxol into the clinic.

Enough taxol was available to begin clinical trials in 1983 – 12,000 trees were needed for around 25,000 kg of bark and this provided 3 – 3.5 kg of taxol. Between 1983 and 1989, the drug was evaluated against breast cancer, small cell lung cancer and ovarian cancer. While there were good responses in the first two cancers (>50% and >20%, respectively), the results in advanced ovarian cancer were particularly exciting. More than 30% of the patients responded to treatment, several long-term remissions being recorded. This may not appear to be startling, but one must remember that whilst ovarian cancer is relatively rare (perhaps one in 70 women will contract the cancer during their lifetime), it can be rapidly fatal unless caught in the early stages. These results led to a certain amount of over-reaction by the American press who labelled taxol as "the most promising anti-cancer drug of the past 15 years", and interest among clinicians became intense.

A major clinical trial was envisaged by the National Cancer Institute, and they planned to treat 12,000 patients per year – a tall order given that an estimated 2 g of drug would be needed for each patient. The commercial and environmental magnitude of the problem of supplying the massive quantity of yew bark required was obvious – the Pacific yew could easily have been eradicated within the space of a few years – and the NCI did not have the resources to seek a solution. In the event, the US pharmaceutical company, Bristol-Myers Squibb, took on the problem in 1991, and as an incentive, they were given exclusive rights to supply the drug to the NCI for its further clinical evaluations. The company would also have exclusive access to the clinical data. Once the trials were complete and the drug had obtained Federal Drug Administration clearance in 1992, the company then had a five-year exclusive right to market the drug on a worldwide basis.

The supply problem was partially solved through improvements in the extraction process from Pacific yew, but a more radical solution was achieved by French groups under the leadership of Pierre Potier in Paris and Andrew Greene in Grenoble. They discovered that the needles of the European yew, *Taxus baccata*, contained a structural relative of taxol called 10-deacetylbaccatin III to the extent of around 1 g per 3 kg of needles. This natural product could be converted to taxol in four chemical steps in a process discovered by Robert Holton in the USA. Since the European yew grows relatively rapidly, the needles could be harvested and the supply of taxol was thus assured.

With the supply lines secured and FDA approval obtained, Bristol-Meyers Squibb were able to secure a unique position in the marketing of this new 'magic bullet'. They even patented the name taxol insisting that their drug be known as Taxol^{R} or paclitaxel. This caused not a little disquiet amongst the chemical community, since it was an unwritten rule that once a natural product had been named and its structure elucidated, no patent could be assigned. The correspondence on this in the pages of *Nature* make fascinating reading.

In France, several new analogues of taxol have been produced by the Potier group, and one of these, Taxotere^{R} (or docetaxel), is at least as active as taxol in clinical trials. Finally, like most natural products, taxol has succumbed to total chemical synthesis, with the US groups of Kyriacos Nicolaou and Robert Holton reporting the results of their efforts almost simultaneously in February 1994. However, these syntheses involve at least 30 discrete stages and are certainly not commercially viable, although they may allow the production of interesting analogues. More recently, Wender, Danishevsky, Kuwajima and Mukaiyama have all reported shorter syntheses, but these are still not commercially viable. Exploration of taxol production by the fungi *Taxomyces andreanae* and a *Periconia* species isolated from *Torreya grandiflora* has been of limited success, and semi-synthesis from 10-deacetylbaccatin III is likely to remain the primary source of taxol and analogues for the forseeable future.

So where does this leave taxol? There is no doubt that it has potent anti-cancer activity and can extend the lives of patients with advanced ovarian and breast cancers in particular, and it is currently undergoing clinical trials in combination with *cis*-platin as a treatment for lung cancer. But its relative water-insolubility and its toxicity to bone-marrow cells make it fall far short of the 'wonder-drug' status that it achieved in the late 1980s. That said, Bristol-Myers Squibb currently has sales of paclitaxel in excess of $1500

Front cover of Nature (17 February 1994) announcing the first chemical synthesis of taxol (Reproduced with permission from Nature, copyright 1994 Macmillan Magazines Limited)

million each year. A large number of analogues have now been prepared and structure–activity studies have revealed the primary requirements for effective anti-cancer activity. These results should facilitate identification of the basic *pharmacophore* (that is, the three-dimensional structure required for

SCIENCE AND LAW

Trademarks must not go generic

Neil White and Simon Cohen

The worldwide registration by a pharmaceutical company of the word *taxol* as a trademark suggests that procedures for registering such marks needs to be tightened.

THE news that the pharmaceutical company Bristol-Myers Squibb has registered the name *taxol* as a trademark in nearly 70 countries worldwide, potentially giving them the right to prevent anyone else from using the name, has given rise to widespread comment in the scientific community[1].

The company is seeking a monopoly on the name, even though it has been used by scientists throughout the world for many years to describe a chemical isolated from the pacific yew tree *Taxus breviofolia*. That product is still undergoing trials. But it appears to help in the treatment of cancer, and Bristol-Myers Squibb wishes to use the name to describe its own proprietary cancer ...

merce. But a genuine trademark, if pr[...] erly policed, can be an invaluable as[...] For example, whereas patent protect[...] lasts for a limited period of time — Europe, this is 20 years from filing trademark protection can potentially for ever, as long as renewal fees are [...] and the mark does not become pa[...] everyday vocabulary.

A trademark need not be a word can be a distinctive shape or get-u[...] even perhaps a smell or sound). Alth[...] under the previous law, the UK Hou[...] Lords refused to register the disti[...] shape of the Coca-Cola bottle, in co[...] they allowed ... registratio[...] ... multi-colour ... Under the ... pected that ... will become ...

Names for hi-jacking

"Taxol" is a trademark now, but Bristol-Myers Squibb should return it to the research community.

EVERYBODY knows what is meant by the English noun "rock". It has three principal uses, in references to stones (that may be thrown), geological deposits (that may be laid down) and a form of rhythmic and cacophonous music, much enjoyed by young people. So what would happen if some recording company were able to proclaim that others should not use the term in any of those meanings on the grounds that "Rock" had become the trademark of its own brand of music? There would, of course, be a revolution. What geologists would do, and to whom, is too dreadful to imagine.

Sadly, there will be no such reaction to the exercise of similar claims by Bristol-Myers Squibb, the US pharmaceutical manufacturer, which has appropriated the word "taxol" for use as a trademark, with the willing but naive consent of the OPT. When *Nature* last year published an account of the total synthesis of this antileukaemia substance (Nicolaou, K. C. *et al. Nature* **367**, 630–634; 1994)., a vice-president of the company wrote to say that "taxol" is a proprietary name and should not be used generically and, that "paclitaxel" is the approved alternative.

For the past two decades, the word "taxol" has been [...] generic name for a material extracted from the bark of th[...] Pacific yew. It was so used in a paper describing the isolati[...] and structure of the compound in 1971 (Wani, M.C. *et al.* *Amer. Chem. Soc.* **93**, 2325–2327; 1971). Bristol-My[...] appears to have inherited its trademark from Contine[...] Laboratories Ltd, which registered the name in 193[...] describe a "pluriglandular product to regulate the i[...] tines". In 1992, Bristol-Myers applied for (and was gra[...] protection of the same name to refer to its "anti-c[...] preparations", making a passing reference to the [...] protection for something quite different.

Bristol-Myers should be ashamed of itself. If it v[...] relationships with the research community, it shoul[...] tarily relinquish the protection it has won "in mor[...] countries". Meanwhile, the US president or o[...] underlings might usefully enquire why the trade[...] aminers were asleep in 1992.

The use of Taxol as a trademark

SIR — I find it, to say the least, ironic that a publication that has adopted the name "Nature" for a magazine about scientific research concerning "nature" should be contending that a trademark of another company is generic (*Nature* 373, 370; 1995): I must assume that Mother Nature and the trademark examiners were "asleep" when you adopted the name "Nature" for your magazine. The aphorism about people who live in glass houses comes quickly to mind.

Pharmaceuticals are a highly regulated field. Trademarks for pharmaceuticals are subject to review by trademark offices and health authorities around the world. Generic names for pharmaceuticals are also subject to regulation by the World Health Organization (WHO), as well as local authorities such as the United States Adopted Names Council (USAN) and the British Pharmacopeia. After review by numerous authorities in numerous countries, Taxol was approved for our use as a trademark for an anticancer preparation. The generic term approved by WHO, USAN and the British Pharmacopoeia for this product is 'paclitaxel.'

Our Taxol anticancer preparation is sold in more than 40 countries under the trademark Taxol and the trademark is now registered in nearly 70 countries. It is well-known as a trademark of Bristol-Myers Squibb throughout the oncology community, as is the generic name paclitaxel. The generic name paclitaxel has also become well-recognized in the research community; there are numerous scientific articles using the approved generic name paclitaxel. The fact that some earlier scientific articles used the term 'taxol' as a trivial name has no bearing on whether Taxol is a recognized trademark among oncologists, which it clearly is. To change the brand name of our product, as you suggest, would cause massive confusion and endanger the health and safety of oncology patients. That is a risk we will certainly not take because of a magazine editorial or indeed for any other reason.

I hope for your sake that Mother Nature doesn't wake up and notice your misappropriation of her name. She's been known to have quite a temper.

Stephen Chesnoff
Bristol-Myers Squibb Company,
345 Park Avenue,
New York, New York 10154–0037, USA

■ *Nature* has never prevented anybody from using the word nature as a common noun; the Bristol-Myers Squibb trademark would end a well-established usage. — Editor, *Nature*.

The correspondence on the trade marking of taxol in the pages of Nature *makes fascinating reading (reproduced with permission from* Nature, *Vol. 373 p. 370, Vol. 373 p. 208, Vol. 375 p. 432, copyright 1994 Macmillan Magazines Limited)* binding of a drug) of taxol and this should allow the design and synthesis of more effective taxol analogues.

MICROORGANISMS ALSO PRODUCE ANTI-TUMOUR AGENTS

When the *Penicillium* mould contaminated Fleming's culture plate in the summer of 1928, it did more than trigger the research effort that ultimately led to the hundreds of antibiotics we use today. It also alerted microbiologists and chemists to the prospect of discovering other biologically interesting constituents of microorganisms. While the moulds have yielded hundreds of highly potent antibacterial agents, the soil microorganisms have yielded literally thousands of exotic natural products with a large variety of pharmacological properties. The *Streptomycetes,* members of the large family of soil microorganisms known as the Actinomycetes, have proved to be a particularly rich source of interesting compounds.

As mentioned in Chapter Two, a pioneer of these investigations was Selman Waksman of the New Jersey Agricultural Experimental Station at Rutgers University. One of his beliefs was that soil microorganisms were responsible for the destruction of pathogenic bacteria when these were present (due to contamination by excreta, *etc*.) in soil. He began his experiments in 1940 by making crude extracts of over 200 different soil samples and almost 50 of these had potent antibacterial activity. The subsequent discovery of streptomycin and other antibiotics has already been related, but it was his later studies on *Streptomyces parvullus* in 1953 that are of particular interest here. This organism was shown to produce a substance they christened actinomycin D – later to be renamed dactinomycin. This had a chemical structure similar to another substance previously isolated from *Streptomyces chrysomallus* by Hans Brockmann of the Organic Chemistry Institute at Gottingen University in 1949. Waksman had the presence of mind to have this new agent tested by Sydney Farber in Boston, and it proved to have potent activity against a range of tumours that had been transplanted into mice. This activity was so marked – the greatest that Farber had seen at that time – that a clinical trial against childhood leukaemia was started almost immediately. Unfortunately, the results were not promising, although subsequent trials showed that the drug was the single agent of choice for effecting a cure of Wilms' tumour (of the kidney) in children, especially if used in conjunction with radiotherapy and other drugs. This tumour is now curable in almost every case (if caught in time) and similar good results have been obtained with Ewing's sarcoma (of bone) and rhabdomycosarcoma (of muscle). The drug's major dose-limiting adverse effect is bone-marrow toxicity (myelosuppression).

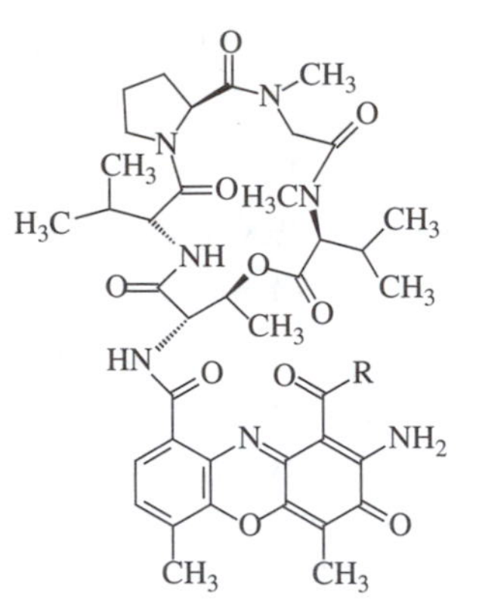

dactinomycin (R = identical cyclic pentapeptide viz: ---threonine-D-valine-proline-N-methylglycine-N-methylvaline----)

Dactinomycin has a very interesting chemical structure that allows it to bind to DNA in two ways. Part of its structure has a disc-like shape and this allows it to slide between the coils of DNA – this is known as *intercalation*. X-ray crystallographic studies have revealed that it associates primarily with base pairs comprising guanine and cytosine through H-bonding. Attached to the central portion of the drug are two chains of amino acids and these bind quite separately in the minor groove of DNA. This tight, co-operative binding inevitably leads to disruption of the replication and transcription of DNA, and there is evidence that the drug inhibits the enzyme topoisomerase II as well. Not surprisingly, this combination of disruptive affects ensures that the drug is one of the most potent anti-cancer agents known.

Historically, mitomycin C was the next anti-tumour agent to be isolated from a soil microorganism. In 1958, Wakaki and co-workers at Kyowa Fermentation Industry in Japan provided the first samples of the natural product from *Streptomyces caespitosus,* while in the USA, researchers at Lederle Laboratories isolated the same compound from *Streptomyces verticillatus,* and after a heroic effort finally elucidated the chemical structure in 1962. It proved to be highly toxic to a range of tumours but also to most kinds of normal cells, resulting in numerous adverse effects when it was used in the clinic. These included nausea, vomiting, bone-marrow depression, alopecia, kidney and liver toxicity and damage to the heart. Nonetheless, the drug became very popular in Japan during the 1960s with around half of all cancer patients receiving the drug as part of their chemotherapy regimens. In the USA, it was produced by Bristol Laboratories and was marketed under the name Mutamycin.

Although mitomycin C has not proved to be a particularly useful drug, it does have a very interesting mode of action. Initially, it must be activated by a process known as *bioreduction*, which results in a number of chemical changes that render it highly reactive towards DNA. It ultimately damages DNA by forming two covalent linkages in a way similar to the alkylating drugs mentioned earlier (Fig. 4.10). Most large solid tumours are hypoxic and have the potential to effect this bioreduction, and mitomycin has found some use in the treatment of such tumours.

A more useful class of drugs are the anthracyclines, of which daunorubicin (daunomycin) and doxorubicin (adriamycin – so named because the microorganism was first isolated on the Adriatic coast of Italy) are the most important natural members. These two were first isolated from *Streptomyces peucetius* in the early 1960s by the Italian group of Arcamone working at the Farmitalia Laboratories in Milan. Literally hundreds of synthetic analogues have since been prepared, and these include marcellomycin, musettamycin and rudolfomycin. Those familiar with the characters in Pucini's opera *La Bohème* will recognise the origin of these names

It was quickly established that adriamycin had broad-spectrum anti-tumour activity and it is still widely used for the treatment of many solid tumours and childhood leukaemia in particular. The other anthracyclines have more limited activities, although daunomycin is the drug of choice (in

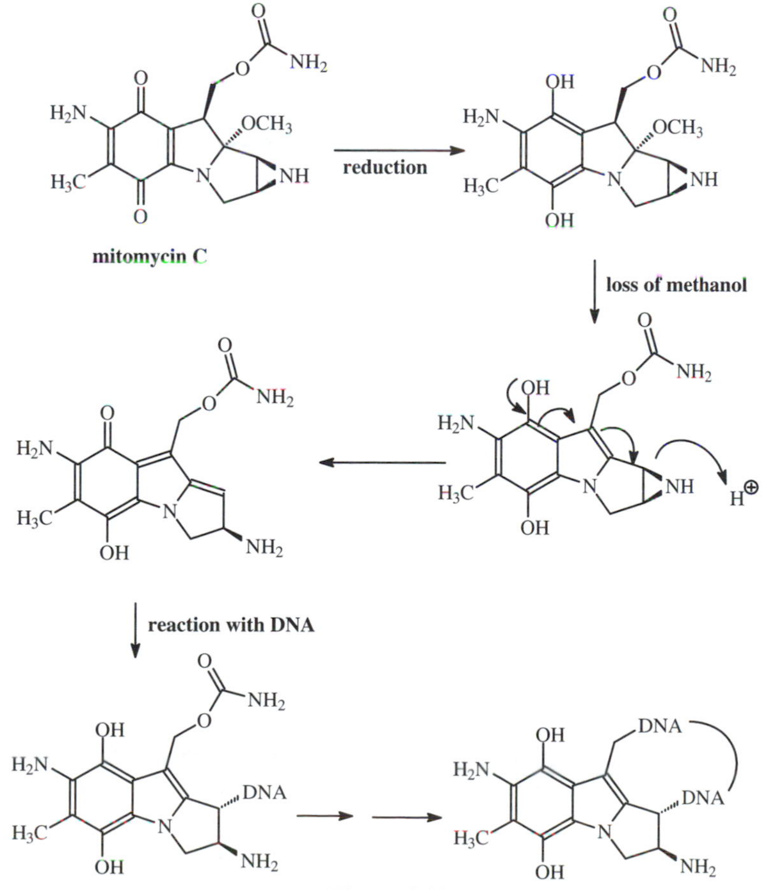

Figure 4.10

conjunction with cytarabine) for the treatment of adult leukaemias. They all function as intercalators because, like dactinomycin, they have a flat, oval shape; but they also appear to interfere with the actions of topoisomerase II. This interference results in DNA strand breaks, which are not subsequently re-annealed. It is also certain that they undergo chemical transformation to produce highly active radicals which react with oxygen to produce superoxide radicals and these damage DNA. The pathway of bioreductive activation to generate these radicals is shown in Fig. 4.11. The superoxide radicals are probably responsible for the main dose-limiting adverse effect, which

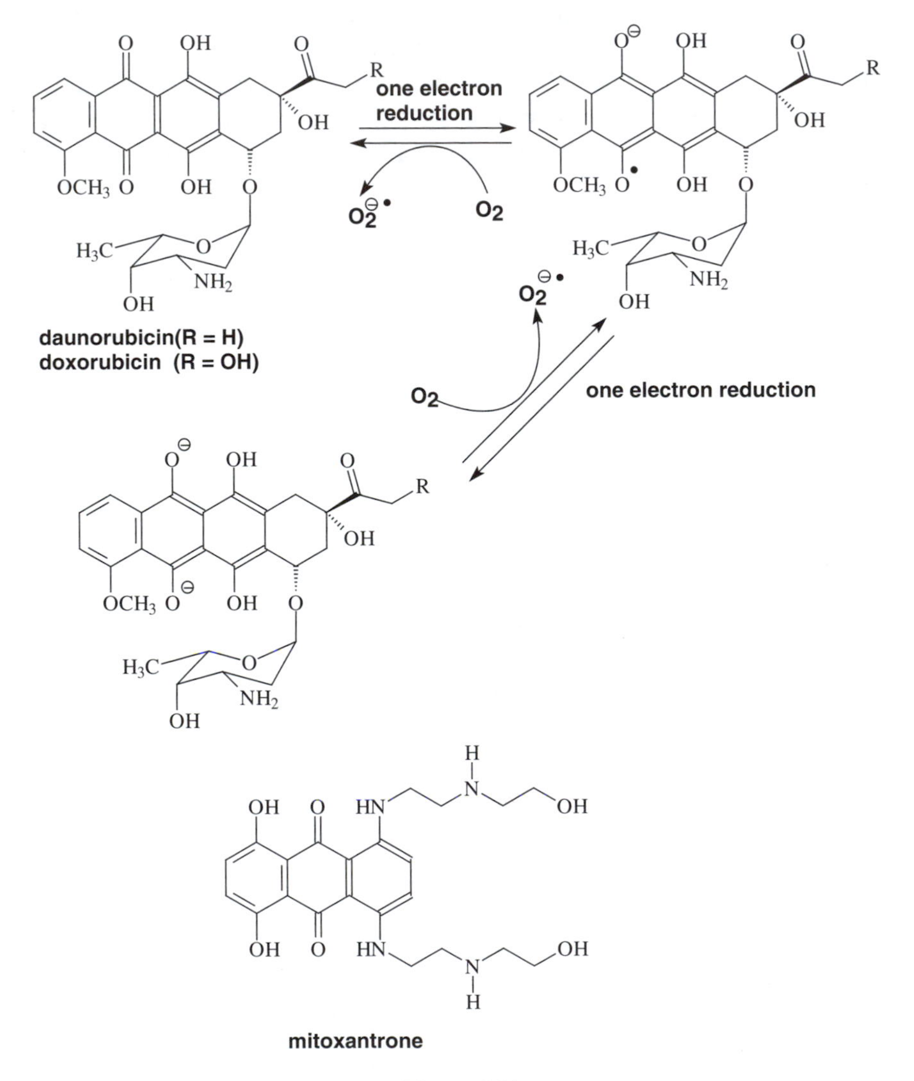

Figure 4.11

involves damage to heart muscle with resultant cardiotoxicity. This toxicity is most marked when patients receive a cumulative dose of more than 600 mg per square metre (of body surface) and at this dosage level, cardiac damage is observed in around 30% of the patients. Many of these suffer from congestive cardiac failure, and in the early days of anthracycline use, many patients died from heart-related problems. One further, totally synthetic analogue of doxorubicin, named mitoxantrone, was introduced in 1978 by Lederle, and has achieved partial acceptance in the clinic. It has less tendency to form radicals and thus causes less cardiotoxicity. It is mainly used for the treatment of breast cancer and leukaemia.

The final class of microorganism-derived products to have made a major impact on cancer chemotherapy are the bleomycins, first isolated in 1966 from *Streptomyces verticillus* by the Japanese group led by Hamao Umezawa. These have a major advantage over the other anti-tumour agents in that they produce little depression of the bone-marrow cell population. Their mode of action also involves cleavage of DNA through the formation of highly destructive radicals. These abstract hydrogen atoms from the DNA strands, which then fragment in a variety of ways (Fig. 4.12). Bleomycin-A2 and B2 are used in combination and comprise the major clinically useful drugs. They are almost always used in combination with other drugs and have activity against a number of tumours including those of the head and neck, cervix and Hodgkin's lymphoma. But its major curative potential has

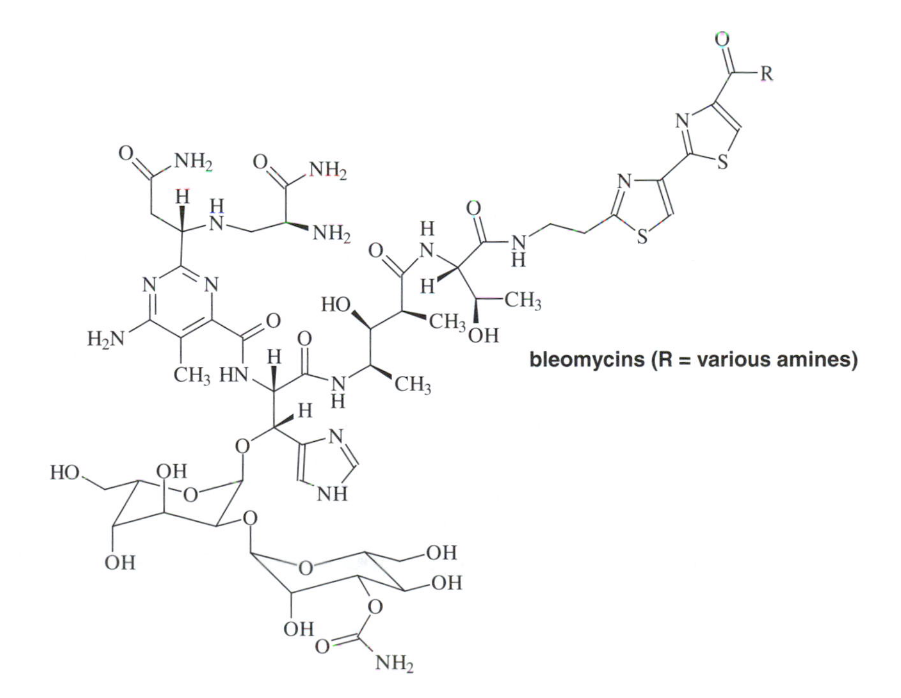

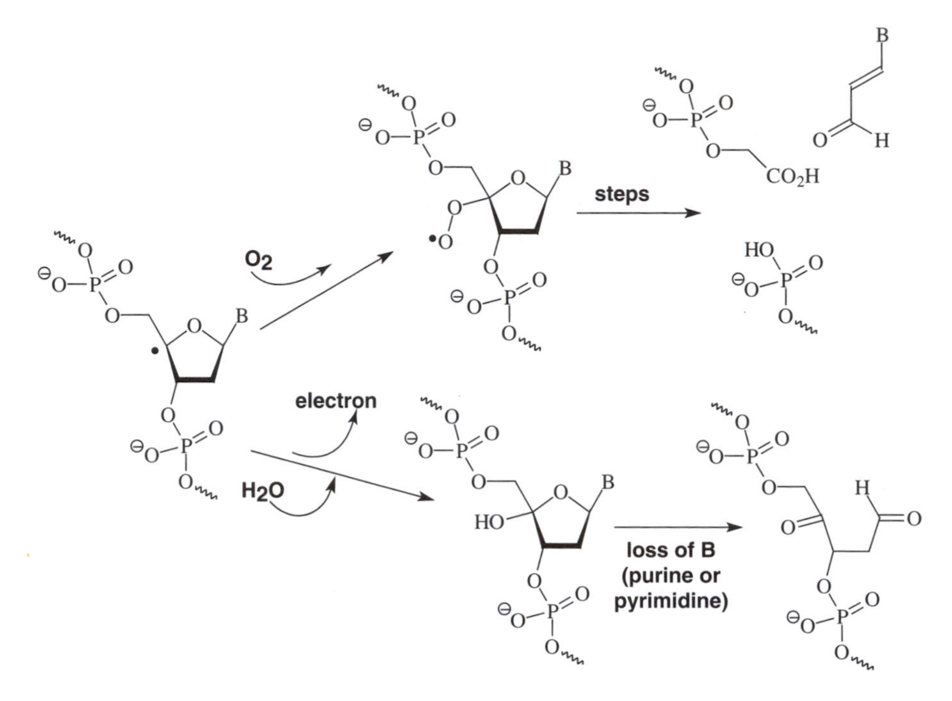

Figure 4.12

been exploited in the treatment of testicular teratoma, in which the response rates increased from 30% (when used alone) to 90% (in conjunction with vinblastine) and to near 100% (with the addition of cisplatin). This combination is usually curative. The drug does produce adverse effects, of which high fever, local tissue damage and lung necrosis can be dose-limiting; but on balance, this is a good drug with thousands of cures to its credit.

Interestingly, it shares its mode of action with another class of anti-tumour agents derived from soil microorganisms – the so-called ene-diynes. With one exception (neocarzinostatin discovered in 1965), these were first discovered in the mid-1980s, and the family now includes almost a dozen compounds with both complex and diverse structures. They are possibly the most exquisitely designed 'magic bullets' that have ever been identified. All of the compounds possess a delivery system usually comprising a sequence of carbohydrates that allows selective binding to the minor groove of DNA (*e.g.*, calicheamicin), or a disc-shaped portion that allows intercalation into the DNA double helix (*e.g.*, neocarzinostatin). Attached to this delivery system is the ene-diyne that can be activated to form a potent benzene diradical, which, like the bleomycins, abstracts hydrogen atoms from DNA and the resultant radicals react with oxygen to produce oxygen-containing radicals, that react destructively with the DNA. In the wild, these ene-diynes certainly prevent the microorganisms from being damaged by invading bacteria or other microorganisms. It is interesting to note that the laboratory version of the ene-diyne to benzene diradical reaction was first discovered by Bergman in 1972, several billion years after Nature first used this process.

In preliminary tests against cancer cells growing in culture, the ene-diynes have shown anti-tumour activity at least as potent as the best clinically used drugs. However, their relative scarcity and potential for damaging normal cells make them unlikely to reach the clinic. They do, nonetheless, represent superb models of agents that are beautifully designed to target DNA, and research is under way to identify the essential structural features that allow this targetting to occur. It may then be possible to design and synthesize simpler chemical structures with more selective anti-cancer activity.

ANTI-TUMOUR DRUGS FROM THE SEA

Viewed from space, it is clear that the Earth is a very watery place, and it is hardly surprising that the oceans support about half of all the life-forms found in the biosphere. Estimates of the total number of different species vary widely and range from around three million species to an upper limit of 500 million species. In the mid-1960s, George Pettit of the Department of Chemistry at Arizona State University in Tempe began his pioneering and systematic investigation of the natural products produced by marine vertebrates and

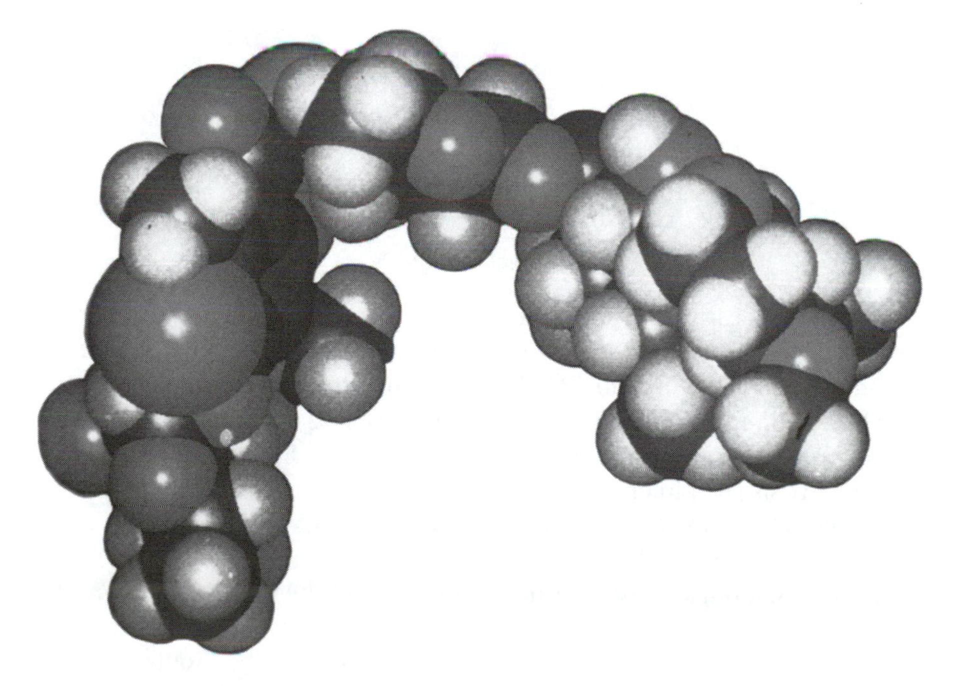

3D structure of calicheamicin—the molecule binds in the minor groove of DNA

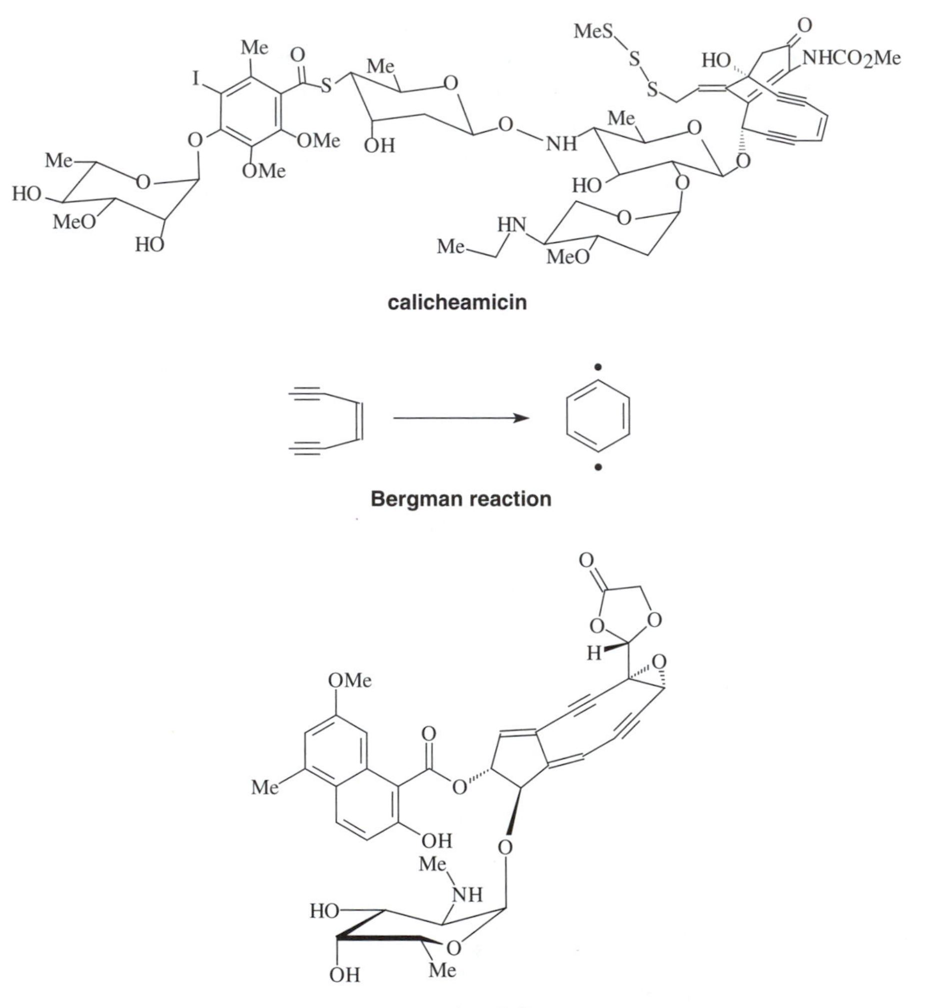

calicheamicin

Bergman reaction

neocarzinostatin

invertebrates. As early as 1969, his group had shown that as many as 10% of the marine organisms they had studied produced compounds that could inhibit the growth of the mouse leukaemia P388. Since then, a growing catalogue of highly complex and in many cases, very potent anti-cancer agents have been discovered by Pettit's group or by other members of a rapidly expanding band of marine chemists and biologists. This work combines the excitement that always accompanies the discovery of new chemical entities with the thrill of diving in (mostly) warm oceans with beautiful (and sometimes dangerous) creatures.

Prominent amongst these marine discoveries are the bryostatins from the marine bryozoan *Bugula neritina*; the dolastatins from the shell-less mollusc *Dolabella auricularia*; the cephalostatins from the South African marine worm (a mere 5 mm long) *Cephalodiscus gilchristi;* the sponge metabolites halichondrin B, the halistatins, dioscodermolide, the spongistatins, the cribrostatins from a beautiful blue sponge found near the Maldives in the Indian Ocean; and the sarcodictyins and eleutherobin. The sarcodictyins were first isolated from a Mediterranean coral *Sarcodictyon roseum* by the group of Pietra in 1987, while the structurally very similar eleutherobin was reported in 1994 by Fenical who isolated the compound from a coral of the *Eleutherobia* species found in the waters off Rottnest Island just offshore from Perth in Western Australia. The groups of Nicolaou and Danishevsky have published elegant, if complicated, total syntheses of these fascinating molecules, and Nicolaou has also produced hundreds of analogues in an attempt to establish structure – activity relationships and define the pharmacophore. Like taxol and discodermolide, the natural products stabilise microtubules during mitosis and thus lead to disruption of cell division. Eleutherobin is claimed to be up to 100 times more potent than taxol in tests with a variety of cancer cell lines, and it is likely that these molecules or simpler analogues will form the basis for effective anti-cancer therapy.

Of these various marine natural products, bryostatin 1 has progressed furthest in clinical trials and has excellent activity against several types of leukaemias and lymphomas. However, the problem of supply is even more acute in this area than with terrestrial organisms. The original sample of cephalostatin 1 (around 140 mg) was obtained from about 170 kg of the marine worms, and given its small size, the magnitude of the task facing the diver is obvious. However, the situation is not completely hopeless, and a small company called CalBioMarine Technologies based in Carlsbad, California, is experimenting with offshore bryozoan farming in order to secure a supply of bryostatin 1. It is also hoped that a genetically modified variant of *Bugula neritina* can be produced that will manufacture large amounts of this interesting anti-tumour agent. For the moment, they have had some success with aquaculture, and typically, a 1 kg sample of farmed *Buluga neritina* will provide about 1.5 g of bryostatin 1 and 2 g of other members of the family. Given the very high potency of these compounds, it has been suggested that as little as 100–200 g of bryostatin 1 could be sufficient for the advanced clinical trials that are presently under way. In addition, Paul Wender of Stanford University has made significant progress in the determination of the basic pharmacophore of the bryostatins through a synthetic approach that has provided much of the core structure of these natural products. Several of his simplified compounds are more potent than their more complex natural product counterparts, and are available in around 20 synthetic steps rather than the 40–60 steps required for the total syntheses that have been accomplished. This is now a common theme of anti-cancer drug design: determination of the basic pharmacophore, then synthesis of

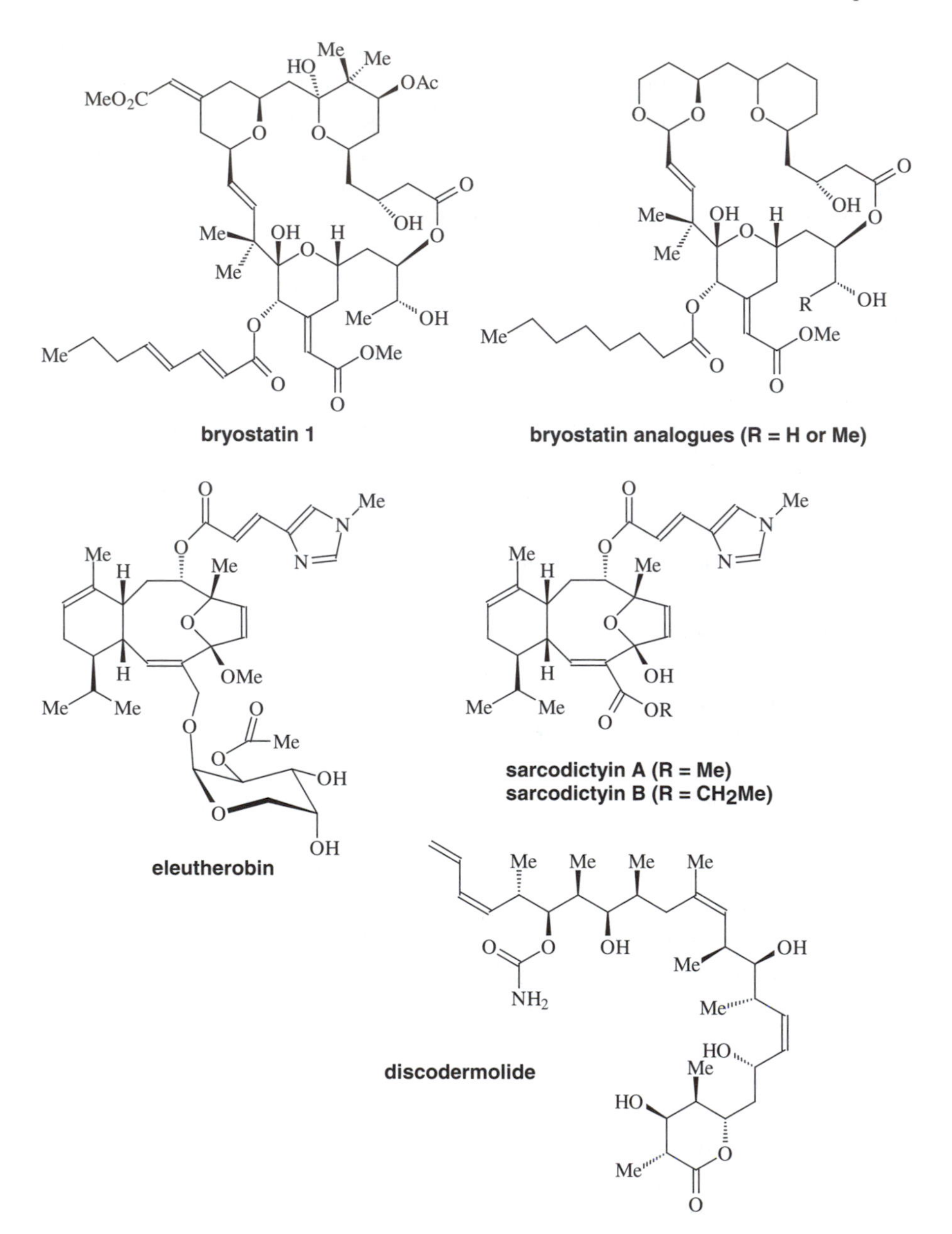

truncated natural product analogues for biological evaluation. The recent total synthesis of both discodermolide and spongistatin on the multigram scale should allow preclinical evaluation of these potentially very exciting marine natural products. For the marine natural product ecteinascidin 743, from the Caribbean sea-squirt *Ecteinascidia turbinata*, a combination of efficient total synthesis by the group of E. J. Corey, and aquaculture by CalBioMarine, has provided sufficient material for advanced clinical trials.

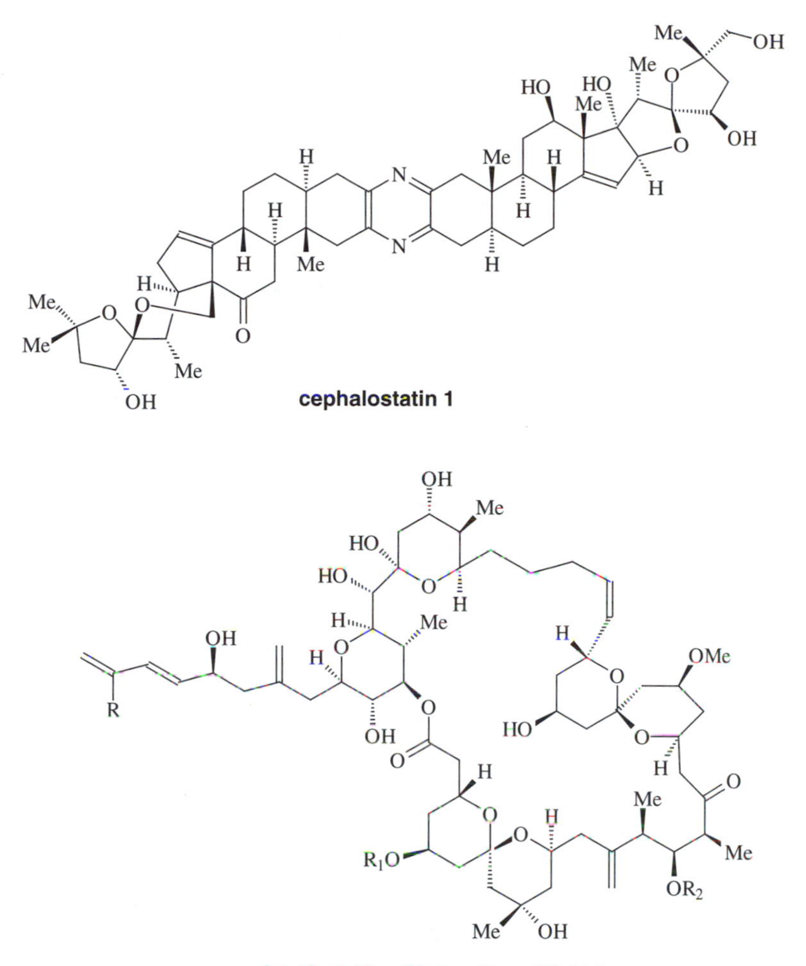

cephalostatin 1

spongistatin 1 (R = Cl, R_1 = R_2 = CO.Me)
spongistatin 2 (R = H, R_1 = R_2 = CO.Me)
spongistatin 3 (R = Cl, R_1 = H, R_2 = CO.Me)

The drug is particularly effective against small cell lung cancer and melanoma, and an analogue synthesised by Corey and given the name phthalascidin, also shows considerable promise.

Finally, returning to natural products of bacterial origin, probably the most promising candidates for clinical use are the epotholones from the myxobacterium *Sorangium cellulosum*. Epothilones A and B were first isolated by Hofle and Reichenbach in 1993 from a soil sample obtained from the banks of the Zambezi river in South Africa. They were tested as anti-fungal agents but were too toxic for human use. Workers at Merck in the USA later demonstrated that these new compounds had a mode of action similar to taxol as stabilisers of microtubules, with potencies that were up to 5000 times greater than taxol

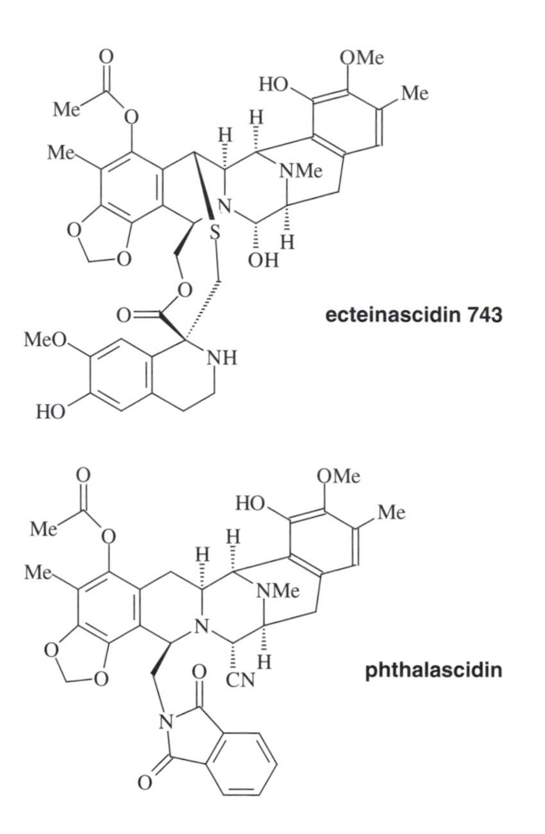

in certain cancer cell lines. A huge amount of synthetic activity has ensued and total syntheses have been reported by Nicolaou, Danishevsky, Schintzer, Furstner and many others. Libraries of analogues have also been reported by Nicolaou, Danishevsky and Hofle, and proposals have been made about the

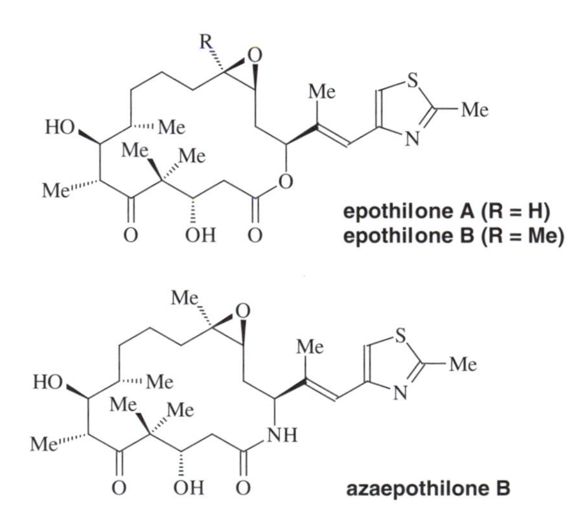

likely pharmacophore for the compounds. Presently, azaepothilone B appears to be the most promising analogue in clinical trials; but what could be most useful about these compounds is that they are not only much more water soluble than taxol, but that they can be produced by fermentation technology. The supply problem that dogs so much of the work with marine anti-cancer agents is thus not a major issue, and these new products from Nature's 'medicine chest' will surely have a very exciting future.

TURNING OFF HORMONAL INFLUENCES

Writing in the *Lancet* of July 18th, 1896, the young Scottish surgeon George Beatson made the seminal observation: "...there are grounds for belief that the aetiology of cancer lies not in the parasitic view but in an ovarian and testicular stimulus..." This view arose from his work at the Glasgow Cancer Hospital on the treatment of women with advanced breast cancer, whose tumours ceased to grow and usually regressed after removal of their ovaries and fallopian tubes. His decision to try this form of treatment was inspired by his knowledge of the interplay between the mammary glands and the ovaries. Although the sex hormones were not isolated and characterized until the 1930s, it was known that substances produced by the ovaries led to development of the breasts during puberty. In contrast, whilst a mother was nursing her baby (that is, during lactation), the ovaries were quiescent in the sense that the menstrual cycle was usually absent. In essence, there appeared to be some kind of chemical communication between these two parts of the body. As he pointed out in his *Lancet* letter: "Above all, I was struck with the local proliferation of epithelium seen in lactation. Here was the very thing characteristic of carcinoma of the breast, and indeed, of the cancerous process everywhere, but differing from it in control by another organ."

His first patient was a young mother of two (aged 33) who had already had one breast removed, yet the cancer was still spreading and her case seemed hopeless. In June 1895, he removed her ovaries and fallopian tubes. Five weeks later the tumour masses had shrunk in size and had become softer to the touch, and seven months later he was able to state: "All vestiges of her previous cancerous disease had disappeared...and she is apparently in excellent health."

Encouraged by this success, he carried out the same operation on a second woman whose case seemed even more hopeless. She was 40 years of age and reported that she had first noticed the growth in her right breast about five years earlier. At the time Beatson first saw her, the tumour extended over much of the breast and there were numerous small tumour masses under her arms and in her neck. Clearly, the cancer had metastasized via the lymphatic system to these various lymph nodes, and she was in considerable pain. She had already been refused treatment at two other local infirmaries on the grounds that her case was hopeless. After the removal of her ovaries and fallopian tubes (on October 3rd) the woman made a remarkable recovery, and by October 14th, she reported that all of her pain had gone and that she was "in a different world." At the time of writing the *Lancet* letter in July 1896, his patient was still in remission.

These highly positive results led Beatson to speculate that certain male cancers, most notably prostatic cancer, might also be under the control of the male secretory organs the testes: "I think that it is possibly in the direction of an altered condition of the ovary and testicles that we are to look for the real exciting cause of cancer." Beatson's speculations have proved to be remarkably perceptive with regard to the two major hormone-dependent cancers – those of the breast and prostate.

Globally, breast cancer is newly diagnosed in more than 500,000 women each year and there are at least 150,000 deaths due to this cancer. Since most of these data emanate from the developed countries, these figures must represent a substantial underestimate. It is the major cause of death in premenopausal women and the single most important cancer of women, accounting for around 20% of all malignancies. Various risk factors have been identified, and these include a family history of the disease (that is, a genetic predisposition – possession of the gene BRC1A is a particular indication of high risk); an early menarche (the age at which regular periods begin); a late menopause; and a late first pregnancy, which confers a greater risk than an absence of pregnancies. Such considerations can be used to provide an estimate that a 35-year-old woman with a strong family history of breast cancer has a 5% chance of developing the disease by the age of 55 and a 1% chance of dying from it. One crumb of comfort is that use of the contraceptive pill (with low oestrogen content) seems to be unrelated to the development of breast cancer, although there is a small positive correlation between the use of oestrogens in hormone replacement therapy and development of the disease.

Prostate cancer is responsible for 50,000 deaths each year in Europe and North America and its incidence is increasing. At present, it is the second most important cancer in men, but by the year 2010, it is destined to overtake lung cancer and become the number one cause of cancer mortality. Interestingly, it is one of the few cancers for which there is no apparent link with smoking.

Before describing the evolution of the various therapies for these two cancers, it is interesting to consider how the sex hormones control normal maturation and the growth of tumours. Overall control of sex hormone production is the responsibility of the hypothalamus, a small gland at the centre of the brain that produces a polypeptide called gonadotropin releasing hormone, GnRH (or luteinizing hormone releasing hormone, LHRH). This acts on another small gland in the brain called the pituitary and stimulates it to produce two additional polypeptides: luteinizing hormone (LH) and follicle stimulating hormone (FSH). These act on the ovaries to stimulate the production of oestrogens and progesterone (the hormone of pregnancy). In men, LH acts on the testes to stimulate the production of testosterone and other male hormones (androgens). If the concentrations of these sex hormones rise above a certain level, the excess hormones in the circulation act on the pituitary and hypothalamus and switch off (inhibit) the production of LH and FSH. This is so-called *feedback inhibition*, quite a common mechanism of control of biochemical systems.

This level of understanding is a far cry from the situation in the 1930s when the first sex hormones were isolated. During the summer of 1929, there was intense competition between the groups of Edward Doisy at St. Louis University School of Medicine in Missouri, USA, and Adolf Butenandt of the University of Gottingen, Germany. Both groups processed large quantities of urine from pregnant women and extracted the hormones with olive oil. Both reported the isolation of a pure crystalline substance that was subsequently called oestrone. Doisy also went on to isolate 12 mg of oestradiol from 4 tons of sows' ovaries! These oestrogens were shown to be steroids and these successes encouraged other researchers, most notably Ernst Laquer of the University of Amsterdam and Adolph Butenandt of Gottingen, to search for the male hormone counterparts. These efforts were rewarded in 1931, when Butenandt managed to isolate 15 mg of androsterone from 25,000 litres of male urine. In 1935, Laqueur isolated 10 mg of testosterone from 100 kg of steer testis tissue. Larger quantities of androsterone were made available in 1934, when Leopold Ruzicka working at the ETH in Zurich, completed a chemical synthesis of androsterone starting from cholesterol, and one year later, his group and that of Butenandt synthesized testosterone. Finally, progesterone was isolated from human corpus luteum (which lines the womb) in 1934 almost simultaneously by groups in New York, Gottingen, Vienna and Basel. If injected into women who habitually miscarried early in pregnancy, this distressing event could be prevented, and the hormone thus came to be called 'progesterone.'

The mass of subsequent chemistry carried out with these steroid hormones has been described on many occasions (see, for example, *Murder, Magic and Medicine*) and led to the discovery of two major classes of drugs: the oral contraceptives and the steroidal anti-inflammatory agents, with a resultant revolution in birth control and the treatment of rheumatoid arthritis, asthma and skin disease.

But how do these ovarian and testicular hormones exert their effects? All cells that interact with these steroid hormones have receptors for them within the cytoplasm. The presence of oestrogen receptors was first demonstrated by Jensen and Jacobsen in 1962. The hormones pass through the cell membrane and then encounter receptors that are specific for each of the classes of hormones: the oestrogens, androgens, progesterone and glucocorticoids

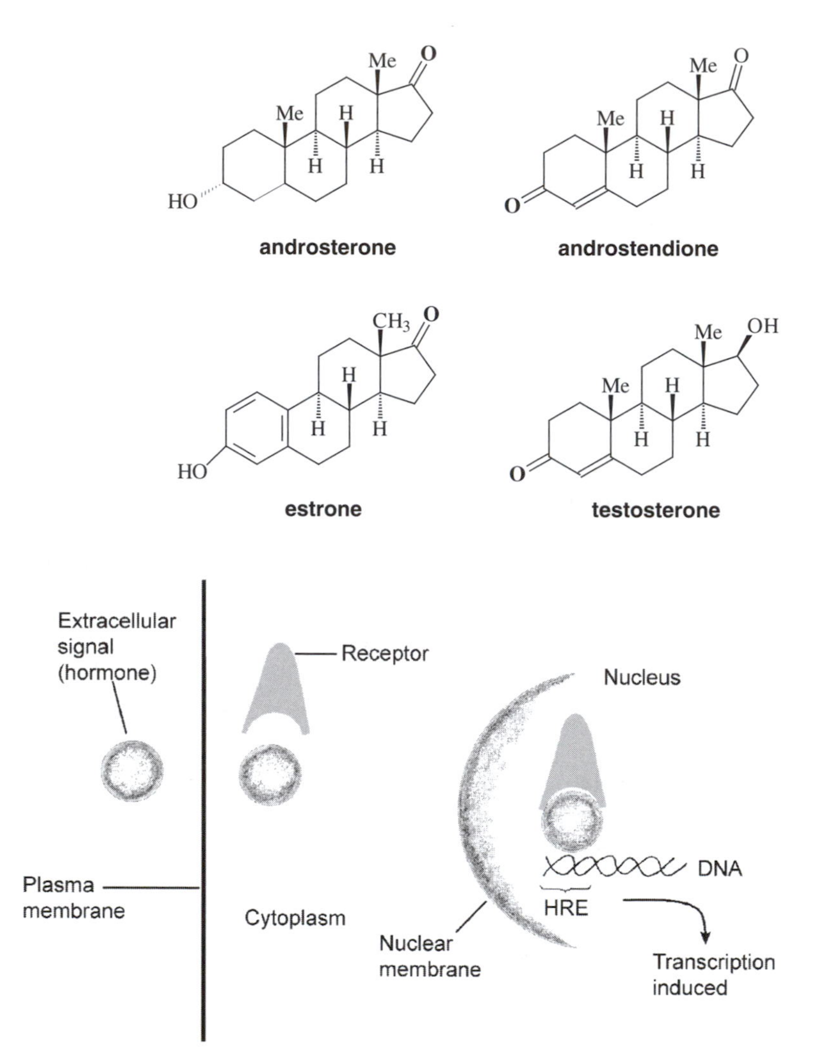

Figure 4.13 *Steroid hormone-receptor interactions*

(steroid hormones that are intimately involved with carbohydrate and protein metabolism). Once associated with the correct receptor, of which there are around 50–60,000 per cell, the hormone–receptor complex is translocated into the cell nucleus, where it interacts with particular sections of DNA known as *hormone response elements*. This nuclear association then triggers a burst of DNA transcription with subsequent production of messenger RNA that specifies the production of proteins that ensure the growth or differentiation of discrete types of cells (Fig. 4.13). The controlled growth and differentiation are clearly hormone-dependent, and it is not difficult to imagine that occasionally (as in a cancer cell), growth might proceed unchecked.

Switching off the production of the hormones or interfering with the formation of the receptor–hormone complexes should thus provide an effective means of therapy. This is indeed the case.

Breast Cancer

The evolution of modern breast cancer therapy can be traced to the pioneering work of Charles Dodds and his colleagues at the Middlesex Hospital in collaboration with the James Cook group at the London Cancer Hospital. They spent most of the 1930s trying to prepare chemical structures that had the basic skeleton of the oestrogens in order to obtain hormone mimics. Several of their compounds did have quite marked oestrogenic activity, and one in particular, a commercial product called anol, appeared to have very potent activity. This was prepared from another commercial compound called anethole, which is a major constituent of the natural plant extract we know as anise. Their initial excitement was somewhat tempered when researchers from other laboratories reported that they could not duplicate the activity with other samples of anol. Even Dodds and Cook found that some of their batches were more active than others, and eventually, the main oestrogenic activity was traced to an impurity in their anol samples. This was subsequently shown to be a dimer of anol. With the help of Robert Robinson of the Dyson Perrins Laboratory in Oxford, they prepared a range of similar dimeric structures, one of which, 4,4′-dihydroxy-α,β-diethylstilbene, was two to three times more active than oestrone itself. In their *Nature* paper of February 5th, 1938, Dodds and Robinson suggested the name 'stilbestrol' for this compound, although this later became known in the USA as diethylstilbestrol or DET. This drug was cheap to manufacture and its high oestrogenic activity, especially when taken by mouth, was a boon to gynaecologists dealing with women who had hormonal problems.

Alexander Haddow of the Chester Beatty Research Institute in London was the first to show, in 1944, that some breast cancer patients could be helped by administration of stilbestrol, and for many years, this was one of the main forms of therapy. Many structural analogues were prepared during the next two decades, culminating in the synthesis of clomiphene in 1959. This stimulated ovulation and was thus used in the treatment of infertility. Some minor chemical manipulation of clomiphene at ICI (now Astra-Zeneca) produced the drug that has become the mainstay of breast cancer therapy today, namely tamoxifen. This exists in two geometric isomeric forms and while one of these had potent oestrogenic activity, the other was a potent anti-oestrogen, that is, it acted by binding to oestrogen receptors in the tumour cells and prevented the attachment of oestrogen, thus effectively preventing the activation of DNA and ultimately inhibiting cell divison in tumours that were oestrogen-dependent.

Tamoxifen (Novaldex) entered a clinical trial at the Christie Hospital in Manchester in 1971 and rapidly became the treatment of choice for post-menopausal patients with breast cancer. Not surprisingly, response rates are typically twice as good for patients with tumours that have oestrogen

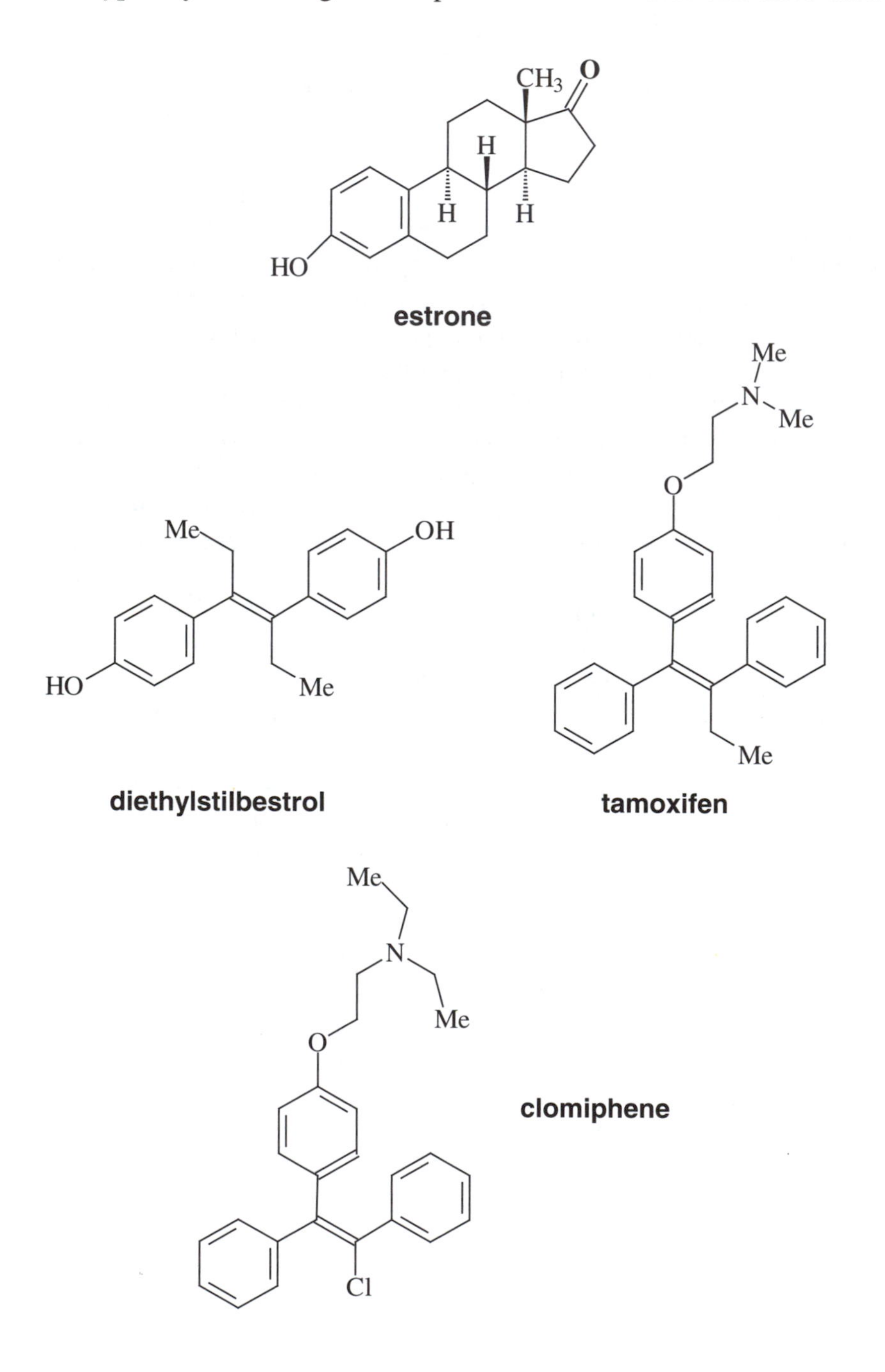

(and also progesterone) receptors, when compared with patients without such receptors. In January 1992, a major report in the *Lancet* described the results of a worldwide collaborative study involving 75,000 breast cancer patients during a 10-year period. Overall, the use of tamoxifen led to a 25% reduction in the recurrence of the disease (following removal of the original tumour) and to a 17% reduction in mortality. It also reduced the risk of tumour appearance in the unaffected breast by 39%. These highly significant results must be considered alongside the other findings of the study. In pre-menopausal women, removal of the ovararies and fallopian tubes is still the main form of therapy, and reductions in recurrence (26%) and mortality (25%) were observed. In addition, other forms of cancer chemotherapy, including the use of a combination of cyclophosphamide, methotrexate and 5-fluorouracil (polychemotherapy), produced a 28% reduction in recurrence and a 16% reduction in mortality. For post-menopausal patients, the combination of tamoxifen and chemotherapy improves risk reduction to 30–40%. And the other major bonus of tamoxifen is that it produces essentially no adverse effects at the usual dose rate of 10 mg twice a day, although there is a slightly increased risk of the development of endometrial cancer and thromboembolism. For the breast cancer patient, the drug does have many of the attributes of a 'magic bullet.' One further potential use for tamoxifen is as a prophylactic treatment for women who have a family history of breast cancer. A major long-term study has commenced and although the final results are not yet available, preliminary results suggest that the drug does reduce the incidence of breast cancer in this group of people.

Another strategy has appeared more recently, and this involves the design of drugs that will inhibit the production of oestrogens in the ovaries by inhibiting the main producer enzyme, aromatase. This enzyme is one of the family known as the cytochrome P450 oxidases, and catalyses the conversion of androstenedione into estrone through the oxidative loss of the C-19 methyl group (see Fig. 4.14). There are a number of drugs, such as aminoglutethimide (Cytadren) and 4-hydroxy-androstenedione (Formestane), that are potent inhibitors of this enzyme, and these are usually administered to patients who have experienced a recurrence following treatment with tamoxifen. More recently, two new aromatase inhibitors, anastrozole and letrozole, have been the subjects of large-scale clinical trials. The results of one large-scale American study on anastrozole were published in 2003. It compared the efficacy of this new drug (Arimidex) versus tamoxifen in a total of 9366 post-menopausal women who had early breast cancer. They were treated in 38 centres in 21 countries between July 1996 and March 2000, and 84% of them had oestrogen-positive and/or progesterone-positive tumours. Almost exactly one-third received anastrozole alone; or tamoxifen alone; the remaining third received a combination of the two drugs. Those receiving anastrozole either

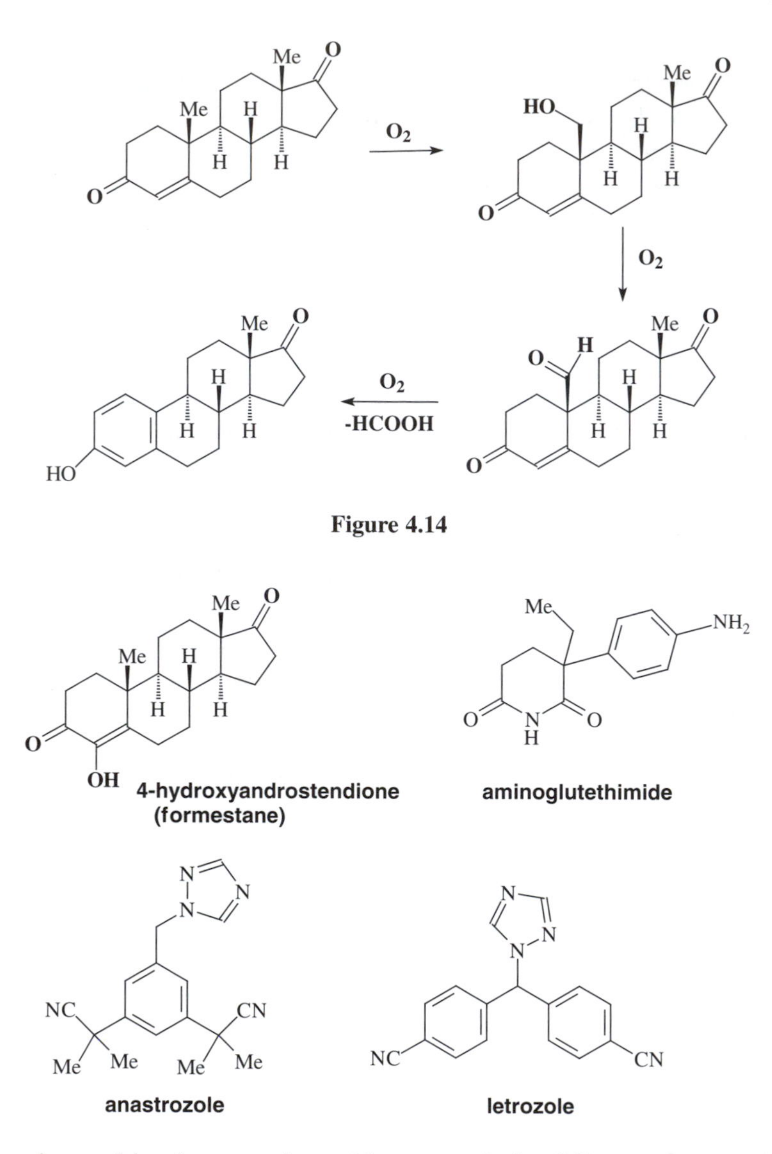

Figure 4.14

alone or in combination experienced longer periods of disease-free survival or longer times before reappearance of their tumours. The incidence of thromoembolism was also lowest in those patients on the anastrozole only treatment regime. The results of a comprehensive European clinical study comparing the efficacy of letrozole (Femara) with anastrozole was also published in 2003. A total of 713 post-menopausal patients with advanced breast cancer, in 112 centres in 19 countries received treatment, and the results appeared to show that letrozole was the most efficacious.

Although it is too early to tell whether the aromatase inhibitors like anastrozole and letrozole will eventually replace the anti-oestrogens like tamoxifen, they do represent a very promising additional treatment for post-menopausal breast cancer.

Prostate Cancer

The prostate is both a blessing and a curse. This small walnut-shaped organ surrounds the urethra (the tube through which urine passes) and is situated at the base of the bladder. It is the source of seminal fluid and is thus intimately involved in maintenance of sexual performance and libido. But to many men over the age of 40 (and to just about all over the age of 70), it is a source of annoyance and discomfort. With advancing age, the prostate grows and begins to constrict the urethra, reducing urine flow – a condition known as prostatic hyperplasia.

That hormones might have an effect on prostate development was hinted at as early as 1786, when the English anatomist John Hunter showed that if young animals were castrated, their prostates failed to grow. But it was the Polish endocrinologist Wugmeister in 1937 who first showed that hormone treatment could have a beneficial effect in his patients with prostatic hyperplasia. He believed that the administration of female hormones would prevent the release of LH from the pituitary (feedback inhibition) with resultant inhibition of testosterone production by the testes. He tested his hypothesis by administering large doses of oestrone to his patients with prostatic hyperplasia with excellent results; but it was two Chicago MDs, Charles Huggins and Clarence Hodges, who first demonstrated the efficacy of female hormones in the treatment of prostate cancer in 1941. First, they discovered that certain serum enzymes, most notably acid phosphatase, were elevated in patients with evidence of prostate cancer, and this was clearly of diagnostic importance. Second, they showed that castration or treatment with oestrogens or stilbestrol caused regression of tumours with concomitant lowering of acid phosphatase activities. For more than 40 years, this became the standard form of treatment for prostate cancer and patients had to suffer the double indignity of 'sacrifice of their manhood' combined with stilbestrol-induced growth of their breasts. Lives were certainly prolonged but the cost was high.

The situation improved with the arrival of the anti-androgens like cyproterone acetate and flutamide, which function like tamoxifen and deny access of testosterone to its receptors in cancer cells. In addition, mimics of the gonadotropin hormone releasing hormone, like leuprolide and goserelin, were discovered, and these act upon the pituitary to switch off the production of LH. This in turn leads to a diminished production of testosterone by the testes. None of these new drugs could effect a cure, although they did cause tumour regression, and where metastasis had led to spread of the tumour to the bones (a frequent problem with this cancer), they provided pain relief. In addition, as with most tumours, the cancer cells eventually become unresponsive to the drugs; hence, other drugs must be sought.

Two recently developed classes of drug show promise, and these act by inhibiting the production of testosterone from cholesterol or its metabolism by the enzyme 5-α-reductase. Attempts to control benign prostatic hyperplasia provided the incentive for the development of 5-α-reductase inhibitors. It has been known for some time that there is a genetic condition that is manifested by a deficiency of this enzyme. Men who have this gene defect have normal external genitalia but only a very small prostate and additionally, they do not develop acne or exhibit the typical male pattern of hair loss. All of these processes are under the control of 5-α-reductase, which controls the conversion of testosterone to another steroid, dihydrotestosterone, and it is an imbalance in the ratio of these two steroids that leads to acne, male-pattern baldness, prostatic hyperplasia and probably prostatic cancer. Several

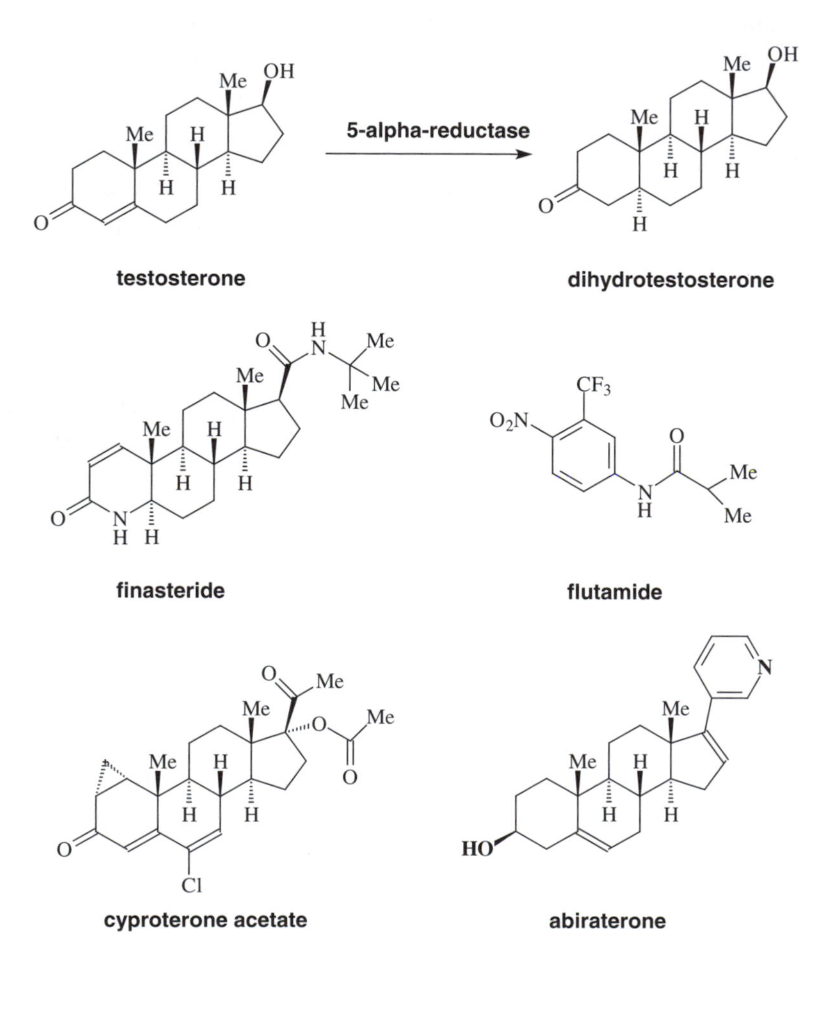

drugs are already on the market for the treatment of prostatic hyperplasia, notably finasteride (Merck), and they are efficacious for this condition. A more dramatic inhibition in the production of testosterone can be achieved through inhibition of the enzyme 17,20-hydroxylase-lyase (another of the cytochrome P450 oxidases), and one of the most active drugs in this area is abiraterone acetate produced by the group of Michael Jarman at the Cancer Research Campaign Laboratory at Sutton.

This is clearly a rapidly developing area and given the growing incidence of this cancer, new drugs are desperately needed. It has been estimated that between the ages of 50 and 75, a man has a 42% chance of developing a slow-growing, essentially undetectable cancer; around a 10% chance of a clinically significant cancer; and a 3% risk of death from prostate cancer. One manifestation of the increasing awareness of the disease is the worrying trend, especially in the USA, of prostate cancer screening. The cancers usually produce what is known as prostate-specific antigen (PSA), which can be detected by a simple blood test. However, although this is a reliable indicator of the presence of tumour cells, it does not determine whether the cells will develop into an aggressive cancer or will simply grow slowly during the lifetime of the patient. In the USA, detection of PSA almost inevitably leads to aggressive treatment regimens, which may not necessarily prolong life but will certainly reduce the quality of life. In the UK, the 'jury is still out' on this issue. An editorial in the *Lancet* (February 15th, 1997) highlighted the report of an NHS committee on this matter which concluded that:

> PSA cannot predict reliably whether a man has a cancer that will progress to cause ill health or death...and routine testing of men to detect prostate cancer should be discouraged.

Clearly, the money spent on costly screening procedures and (possibly) unnecessary surgery and chemotherapy could be better spent on research to discover better drugs to treat this major disease.

BIOLOGICAL WARFARE AGAINST CANCER CELLS

In 1893, William Coley, a surgeon at Memorial Hospital in New York City, deliberately injected a mixture of heat-treated bacterial cells into the tumours of his cancer patients. In all, he treated over 900 patients, and there were several dramatic remissions. He was inspired to do this by the observation that cancer patients sometimes experienced remarkable improvement in their condition following a serious bacterial infection. It was as if their already severely damaged bodies were somehow activated by the bacterial onslaught to mount a sustained attack on both the bacterial and cancer cells. 'Coley's toxins' – usually a mixture of killed bacteria of the species *Streptococcus pyogenes* and *Serratia marcescens* – became very popular and were the mainstay of cancer treatment before the introduction of radiotherapy in the 1930s. Other microorganisms were also tried and these included certain Gram-negative bacteria like Bacillus Calmette–Guerin or BCG – the non-pathogenic form of the tubercle bacillus (used since the Second World War for vaccination against TB) and *Corynebacterium parvum*. Although there were spectacular successes with tumours in mice, the results with human cancers were disappointing. Interest was rekindled when it was discovered that it was lipopolysaccharide or LPS produced by the Gram-negative bacteria that was activating the immune system. Indeed, overstimulation was associated with the so-called 'toxic shock syndrome' (discussed in Chapter Two).

In 1975, Lloyd Old and his co-workers in New York reported that these agents led to activation of the white cells known as macrophages and these are stimulated to produce a number of highly potent cytotoxic factors, most notably *tumour necrosis factor* or TNF. This kills cancer cells by a variety of mechanisms that include formation of oxygen radicals with subsequent fragmentation of DNA, and also activation of the cascade of events that is known as programmed cell death or apoptosis. TNF has a high degree of specificity for cancer cells. However, it is not totally specific and its great potency means that adverse effects can be very serious, and the limited clinical trials with intravenous TNF have been disappointing.

A range of other factors produced by white blood cells have been tried in the clinic and these include the *interferons* (mentioned in Chapter Three) and the *interleukins*. The interferons are produced by activated T-lymphocytes and leucocytes and apparently exert their activity through stimulation of a population of aptly named *natural killer cells* and macrophages. Some limited success has been achieved in clinical trials with interferons, most importantly against a rare form of leukaemia – 'hairy cell' leukaemia (so named because of the appearance of the cell surface) – and Kaposi's sarcoma, another rare cancer but prevalent in AIDS patients.

The interleukins are, like the interferons, polypetides and are produced by lymphocytes, macrophages and other white blood cells. They are intimately involved in the activation of lymphocytes of the T and B classes, which are primarily responsible for a vigorous immune response. At least 18 discrete interkeukins have now been identified and purified, but once again, clinical trials have been disappointing, with serious adverse effects that include the production of fluid in the lungs, fever, nausea, serious vomiting and diarrhoea, *etc*., although some regression of tumours was evident.

But all of these factors have broad-spectrum cytotoxicity and it is doubtful if it will be possible to increase their selectivity towards cancer cells. It would be preferrable if one could switch off the growth and metastasis of cancer cells. Since many cancers are now known to be under the influence of oncogenes, it would be highly desirable if these permanently activated genes could be deactivated. Alternatively, since these oncogenes code for the production of aberrant proteins that allow the cells to grow out of control, deactivation of these proteins could also be a useful strategy. One highly subtle means of control is feasible and this involves the inhibition of a particular enzyme – *ras*-farnesyl transferase. *Ras*-oncogenes are present in around 20–30% of tumours and code for the production of *Ras*-proteins, which are aberrant forms of the normal *Ras*-G proteins. As already mentioned, these are at the head of a complex signalling pathway that normally initiates a cascade of chemical changes which instruct the cell to differentiate or proliferate (Fig. 4.5). Before these *Ras*-G proteins can function, they require the addition of a hydrocarbon tail (a 15-carbon farnesyl moiety) in order to anchor them to the inner surface of the cell membrane. Typically, the enzyme adds the farnesyl residue to a cysteine that is close to the carboxy terminus of the *Ras*-protein. A further addition of a palmitoyl (an 18-carbon fatty acid) unit to another cysteine residue close to the farnesylated-cysteine then ensures that the *Ras*-protein can be anchored to the cell membrane via these two lipophilic tails. Inhibitors of the various enzymes that catalyse the addition of the farnesyl unit should attenuate cancer cell growth, and a few compounds have been evaluated in the clinic. However, the results to date have been disappointing.

At the other end of the cell signalling cascade are the cyclins and cyclin-dependent kinases that control the passage of the cell cycle between the G_2 phase and mitosis. The natural product rohitukine from the bacterium *Dysoxyylum binectariferum* acted as the model for the totally synthetic compound flavopiridol, which has a potent inhibitory effect on cyclin-dependent kinases and is under clinical evaluation. The unrelated naturally occurring purines olomoucine and roscovitine similarly served as 'natural product leads' for the design of the synthetic purine analogues purvalanols A and B. All of these compounds are potent CDK inhibitors, with the purvalanols being around 1000 times more potent than the natural products. X-ray crystallographic analysis has shown that these species compete with ATP for the ATP-binding site on CDKs, and these discoveries provide a further example of how chemists can raid Nature's 'medicine chest' for inspiration when designing drugs for new targets.

A more high-technology approach involves therapy to try to target those cells that have a deficiency in p53 tumour suppressor genes – and that is around 50% of cancer cells. These genes prevent normal cells from making copies of damaged DNA or of foreign DNA (like that of a virus, for example) that might have entered the cell. Certain types of viruses, especially adenoviruses, can apparently inhibit the activities of the p53 genes thus allowing incorporation of the viral DNA into the host genome. In order to take advantage of this mechanism for cancer therapy, it was necessary to produce a mutant adenovirus that lacked the capability to inhibit the suppressor genes.

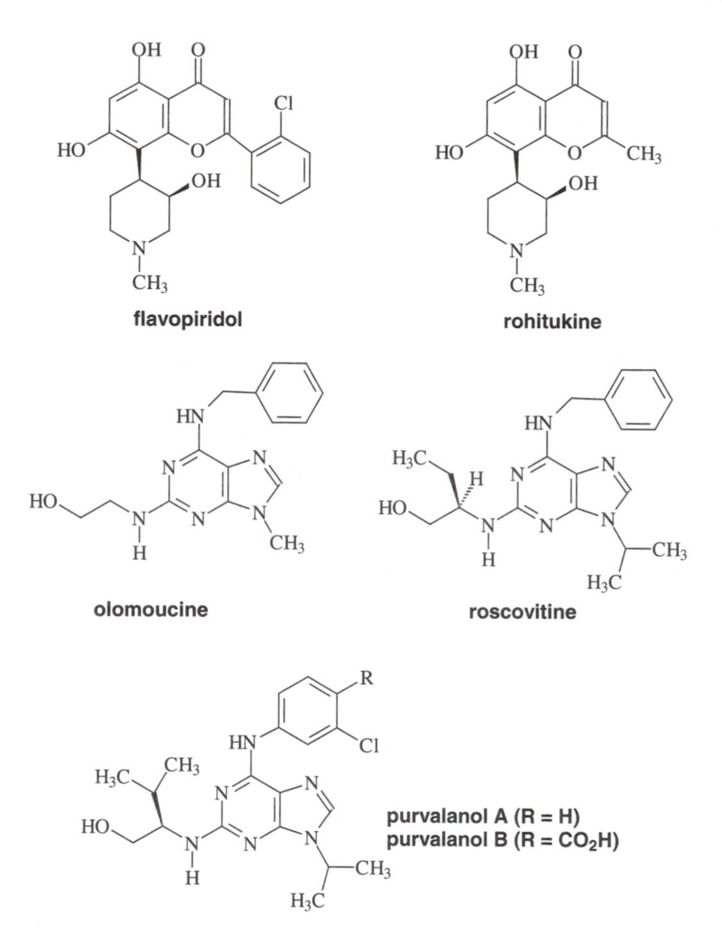

These aberrant viruses could no longer survive and reproduce in normal cells but could do so in cancer cells that lacked p53 suppressor genes. The mutant adenovirus should thus only destroy such cancer cells and a group at ONYX Pharmaceuticals in Richmond, California, have demonstrated the feasibility of this strategy. There is a flaw, however, and this is due to the fact that most adults have encountered adenoviruses and their immune systems are thus capable of eliminating them before they reach the tumour. Thus, the mutant viral preparation has to be injected directly into the tumour and may only have efficacy against localized tumours in the brain, ovaries, head and neck; but this new strategy may be the forerunner of other similarly subtle forms of therapy. Very recently, the first small molecule reactivators for deactivated p53 have been announced. This group of compounds – the nutlins – were part of a large library of compounds that were screened for biological activity, and the study of their interactions with p53 is clearly an exciting area of investigation.

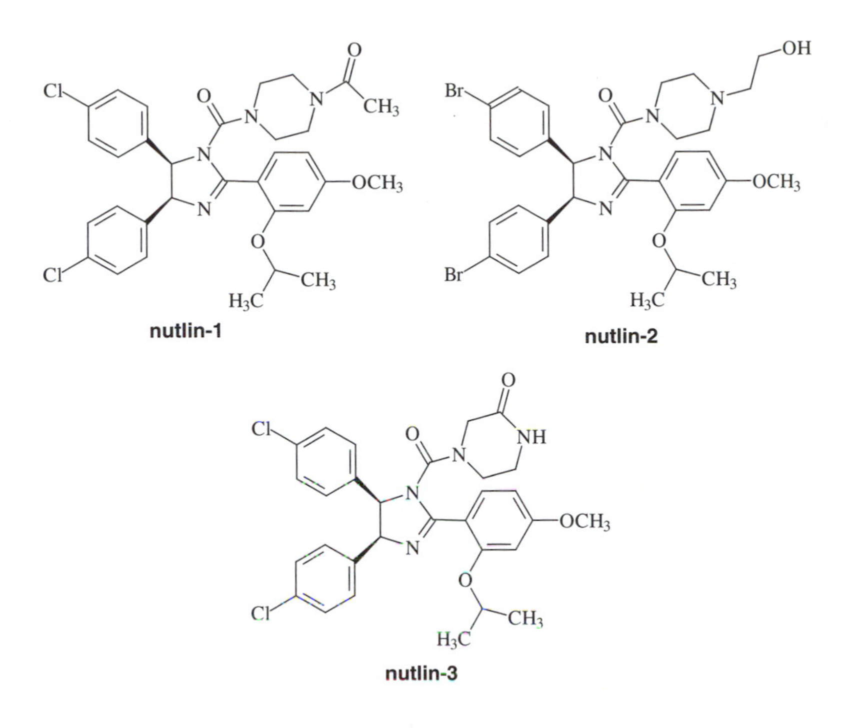

Despite all these advances in drug design, a major problem of cancer therapy is the need to target the drugs to the site of the cancer. Ultimately, what kills the cancer patient is the metastatic potential of the tumour, with its spread to distant organs, especially the brain, lungs and liver. While the tumour is localized, there is a chance of removing it by surgery or destroying it by means of radiotherapy or a form of local drug therapy. One well-tried and tested treatment is known as *photodynamic therapy*, and involves the use of singlet oxygen, produced when oxygen gas is irradiated in the presence of activating chemicals, most commonly porphyrins. A good example of the use of this strategy is seen in the therapy of bladder cancer. A fine catheter is passed via the urinary tract into the bladder and the porphyrin is then introduced together with a stream of oxygen. A fibre optic device is then threaded down the catheter to irradiate the mixture of porphyrin and oxygen in the bladder. This irradiation produces molecules of singlet oxygen, and these damage the cell membrane constituents and the DNA of the tumour cells, without damaging tissues outside the confines of the bladder. More recently, trials have begun to assess the application of photodynamic therapy for the treatment of solid tumours like those of the head, neck and oesophagus. In this case, the porphyrin is administered to the patient via the oxygenated blood in the circulation, and the solid tumour is then irradiated, through the skin, using a red laser light.

The ability to prevent metastasis is another major aim of modern drug therapy and one that has a realistic chance of success. In order to escape the confines of the primary tumour, cancer cells must reach the circulatory system for blood or the lymphatic system. It achieves this goal through the development of new blood vessels, a process known as *angiogenesis*. This requires the combined activities of a number of enzymes, most importantly the matrix metalloproteinases. These are normally involved in the controlled breakdown of the glycoprotein matrix that surrounds most cells during rapid cell growth. Such growth is prevalent during embryonic development and wound healing, but also occurs with populations of cancer cells. The enzymes are required to facilitate the destruction of the extracellular matrix of a tumour mass and allow the tumour cells to enter the bloodstream. Several new drugs that inhibit the activities of these enzymes are currently undergoing clinical evaluation, and the drug marimastat has probably received the most attention. In addition, endogeneous proteins like angiostatin and endostatin, have very potent inhibitory activity against angiogenesis. Angiostatin and endostatin are breakdown products of larger endogenous proteins – collagen and plasminogen, respectively – which are intimately involved in wound healing, hence

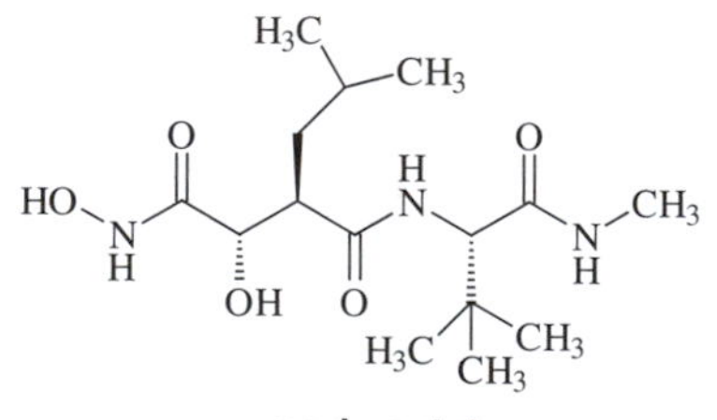

marimastat

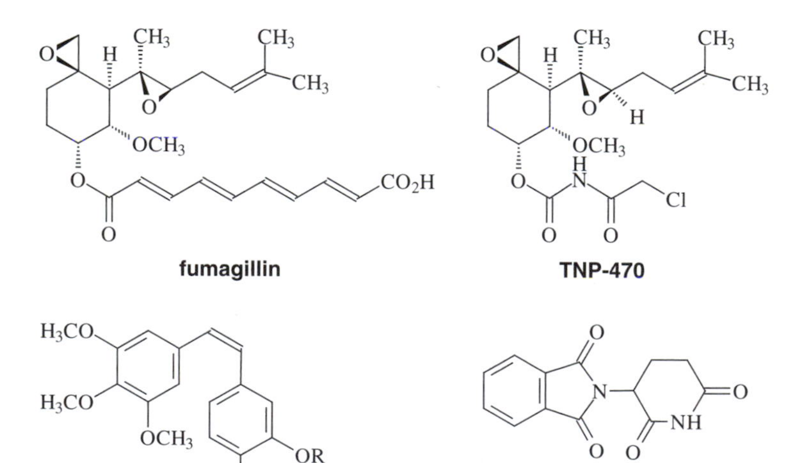

fumagillin

TNP-470

thalidomide

combretastatin A-4 (R = H)
combretastain A-4 phosphate (R = Na_2PO_3)

probably have a normal role in switching off angiogenesis once a wound has healed. They are currently undergoing clinical evaluation against a number of cancers. Two natural products have also shown significant activity as angiogenesis inhibitors – fumagillin from *Aspergillus fumigatus* and combretastatin A-4 from the African bush willow *Combretum caffrum*. Several analogues of these compounds have been screened for improved activity, and TNP-470 (chloroacetylcarbamyl fumagillol) and the water-soluble phosphate derivative of combretastatin A-4 are showing particular promise. Even the justly maligned drug thalidomide has been reinvented as an effective inhibitor of angiogenesis and has shown considerable promise in early clinical trials.

Another exciting possibility arises from studies on *telomeres*. These are lengths of single-stranded DNA added to the ends of newly synthesised chromosomes under the influence of enzymes known as telomerases. These DNA sequences are not part of the actual genetic blueprint, but mark the termini of the chromosomes. During foetal development, when much cell proliferation is occurring, telomerase activity is high; but once the cells have matured, the enzymes are usually absent. As cells age, the telomeres are gradually reduced in length, and eventually the chromosome becomes unstable and the cell must be disposed of by, for example, apoptosis (controlled cell death). In almost all cancer cell types that have been examined (certainly greater than 90%), the telomerase activity is high; hence, the length of the telomeres is maintained. This is believed to confer immortality on the cells. Inhibitors of this enzyme are thus of considerable interest, and some progress has been made in the design of drugs that have this activity.

Perhaps the closest we can come to a 'magic bullet' is to design drugs that are targeted specifically to cancer cells. This became possible following the discovery of *monoclonal antibodies*. In 1975, Cesar Millstein and Georges Kohler, working at the Medical Research Council Laboratory in Cambridge, made the seminal discovery that fusion of an activated lymphocyte with a malignant myeloma cell produced a hybrid cell (hybridoma) that was not only immortal but would also produce large amounts of antibody to the antigen that had initially stimulated the lymphocyte. In hindsight, this is not surprising. Myeloma cells are cancer cells that are programmed to produce antibodies of one particular type and essentially nothing else, and like most cancer cells they are immortal. The hybridoma combines these two attributes, but now produces antibodies against a new antigen, which is in principle open to selection. The research worker can choose the antigen and a population of monoclonal antibodies will be produced by the hybridoma. Of course, the situation is not quite so simple and there are other 'tricks of the trade' that must be used to obtain an efficient population of hybridoma cells; but nonetheless, this technology allows the production of large quantities of highly specific antibodies in a relatively inexpensive manner.

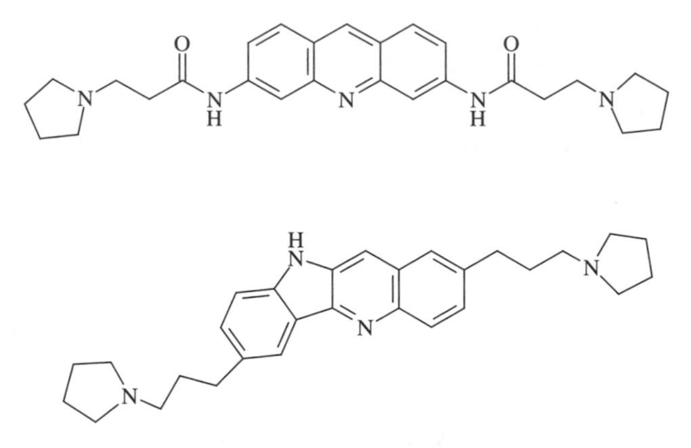

experimental telomerase inhibitors

There has been limited success with monoclonal antibodies, used by themselves, and the antibody product Herceptin is employed in the treatment of certain forms of breast cancer. This antibody targets a protein called ERBB2, which is over-expressed in many breast tumours, and has a normal role as a receptor for endothelial growth factor and thus acts as an upstream activator of *Ras*-G proteins. Monoclonal antibodies have also been used to eradicate small populations of cancer cells from bone-marrow cells that have been pre-

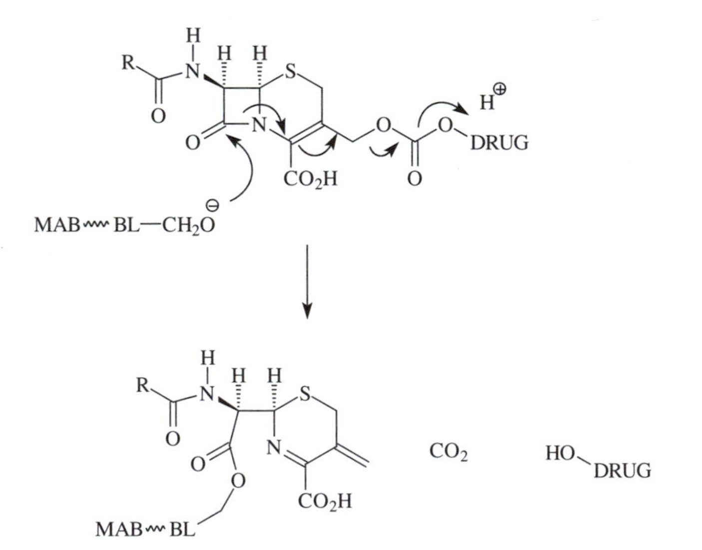

ADEPT: an antibody-beta-lactamase conjugate attacks a cephalopsorin-anti-cancer drug conjugate to liberate the anti-cancer drug

Drugs have included N-mustards, doxorubicin, taxol, mitomycin, vinblastine, etc.

Figure 4.15

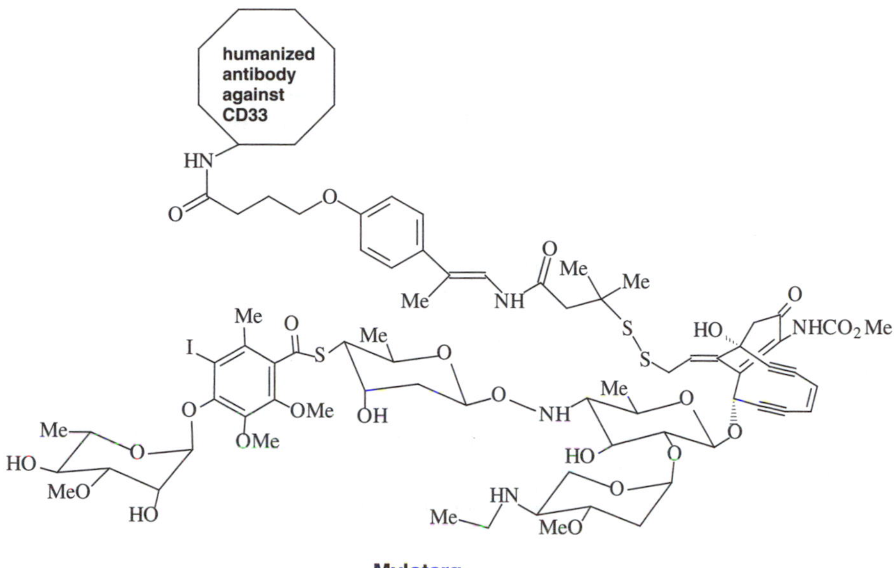

Mylotarg

treated with standard forms of chemotherapy; but their main use has been in the targeting of drugs to cancer cells.

If the monoclonal antibody has been designed to have specificity for a particular type of cancer cell, which of course has its own antigenic 'signature,' it will seek out this cell and become bound to the surface of the cell. Two modes of therapy are then feasible. If the monoclonal antibody has an anti-cancer drug attached to it, this will now be released in the vicinity of the cancer or the whole antibody–anti-cancer drug conjugate may be taken up by the cell. Drugs like taxol and vinblastine, as well as radioactive isotopes and natural toxins like ricin (from the castor oil plant), have been targeted in this way. One of the best examples of this type of drug is Celltech's gemtuzumab ozogamicin (Mylotarg), which is a conjugate of a humanized monoclonal antibody against the CD33 antigen of leukaemia cells with the ene-diyne natural product calicheamicin.

An alternative strategy known as ADEPT (antibody-directed enzyme prodrug therapy) involves an antibody–enzyme conjugate which is targeted at the tumour cell. Subsequently, the patient is given an anti-cancer drug that needs activation (a prodrug) by the enzyme that is already held in close proximity to the cancer cell. As the prodrug passes the cell, the enzyme acts upon it and the resultant active drug is then taken up by the cancer cell (see Fig. 4.15). A successful application of this strategy involves a prodrug comprising a beta-lactam antibiotic attached to the anti-tumour drug vinblastine. In this instance, the enzyme attached to the tumour-specific monoclonal antibody is a beta-lactamase (not an endogenous human enzyme). As the prodrug reaches the tumour cell/monoclonal antibody/beta-lactamase complex, the enzyme attacks the beta-lactam ring of the antibiotic, and as the molecule collapses it releases the vinblastine to be taken up by the tumour cell.

An alternative method of drug delivery has recently reached the commercial stage, and this involves the use of a polymer–drug conjugate. The idea is that the drug is essentially inactive and non-metabolizable while it is attached to the polymer; but once taken up by the cancer cell, the enzymes within the cell remove the polymer, thus revealing the drug. The special requirements of tumour cells for their own blood supply and discrete blood vessels (neovasculature) provide a degree of selectivity, especially since this neovasculature is often leaky to large molecules. Several commercial products are under clinical evaluation, which involve conjugates of polyethylene glycol and known anti-cancer drugs like daunorubicin and the anti-leukaemia drug asparaginase. Another type is based on a modified polystyrene–neocarzinostatin conjugate.

A further novel strategy involves the insertion of thin wafers made up of a biodegradable polymer and the anti-tumour drug carmustine (BCNU), during an operation to remove a brain tumour. The drug is then released slowly over a period of a month or so in the vicinity of any residual tumour cells. In a recent clinical trial in the USA, 31% of the patients who had been treated in this way were alive two years after their operation, but only 6% of a matched group who had undergone the same operation but received no BCNU inserts survived.

A more ambitious therapy involves insertion of a particular gene into the cancer cell. The genes are chosen to be ones that code for the production of particular enzymes that will activate pro-anti-cancer drugs when these are absorbed by the cancer cell. The trick is to smuggle the genes into the cancer cell by means of a virus. This strategy is possible because these pathogenic organisms operate by usurping the biochemical machinery of the cell with the ultimate aim of incorporating their genetic material into the host cell's DNA. Clearly, this strategy, known as GDEPT (gene-directed enzyme prodrug therapy) will be harder to achieve, since the genes must be smuggled into cancer cells rather than normal cells.

Finally, vaccination against cancer is beginning to emerge as a realistic possibility. It will be recalled that Coley's toxins were used to evoke an immune response in an early form of cancer treatment. Now the technique has been reinvented through the introduction of synthetic vaccines raised against portions of the glycoprotein coat of tumour cells. In phase II clinical trials carried out in Canada, the USA and the UK, more than 200 patients with colorectal cancer were treated with this type of vaccine. Most of them experienced the same periods of survival as patients treated with chemotherapy, but their quality of life was better, since they did not suffer from any adverse effects due to drugs.

A synthetic vaccine has been raised against the coat glycoprotein of HPV16 – one of the most oncogenic papilloma viruses that are implicated in the causation of cervical cancer. In a study, 768 women aged 16 to 23 were given three doses of the vaccine while the control group received no vaccination. After 18 months, the control group had suffered 41 cases of infection with HPV16 while the treated group had no infections, thus establishing the preventative efficacy of the vaccine. This is a major step forward since there are at least 450,000 cases of cervical cancer reported each year worldwide, and it is the third largest cause of female cancer deaths after lung and breast cancer. Synthetic vaccines to oncogenes and tumour suppressor genes are also under evaluation.

Many of these potential therapies seem to border on science fiction, but when one recalls that it is only 40 years since the first effective anti-cancer drugs were discovered, it seems highly probable that some of these therapies will be the 'magic bullets' of tomorrow.

PREVENTION IS BETTER THAN CURE

While it is often possible to avoid infection by bacteria and viruses, the means of avoiding cancer seem less obvious. However, given that perhaps 70% of all cancers are due (at least in part) to environmental factors – diet, pollution, smoking and sunlight – it should be feasible to take appropriate precautions that would lessen the risk of cancer. It would be possible to prevent most forms of lung cancer by the simple expedient of banning cigarette sales. The salutary effects on heart disease and several other forms of cancer would also be marked. A reduction in exposure to sunlight, especially during the first decades of life, would also reduce the incidence of skin cancers very dramatically.

We worry with some justification about chemical carcinogens in the workplace – blue asbestos and chromium salts being prime examples – but there is a high degree of over-reaction to things like pesticide residues and artificial flavours and colours in our diet. As the American biochemist Bruce Ames (inventor of the so-called Ames test for mutagenicity) has pointed out:

> There are more natural carcinogens by weight in a cup of coffee than potentially carcinogenic synthetic pesticide residues in the average US diet in a year, and there are still a thousand known chemicals in roasted coffee that have not been tested.

Infection by certain viruses is known to be involved in the induction of certain tumours. For example, the high incidence of liver cancer in Africa and Asia is linked with the prevalence of hepatitis B and C virus infection in these areas, and a programme of mass vaccination could provide a partial solution.

There is no question that our DNA is under assault all the time. The normal biochemistry of life uses vast quantities of oxygen and inevitably, there is much production of highly dangerous oxygen free radicals. We have enzymes that can deal with these species, but it is estimated nonetheless that our DNA suffers perhaps 10,000 damaging events per cell every day. We also have repair enzymes that will cut out the damaged DNA and ensure that replacement is effectively achieved; but we can take steps to minimize this damage. The single most important factor in our daily lives is our diet. A low-fat, high-fibre diet that is enriched with fresh vegetables (especially carrots, cauliflower and broccoli) and fruit will help to ensure good health. It has now been established beyond reasonable doubt that vitamins A, C and E together with other antioxidants (found mainly in fresh fruit and vegetables but also in such unlikely places as red wine and green tea) make a major contribution to cancer prevention by intercepting and destroying cytotoxic oxygen radicals. In crude terms, that section of the population with the lowest intake of these foods has about twice the incidence of cancer as the section of the population that consumes the largest quantity of fruit and vegetables. Unfortunately, the link between diet and the hormone-induced cancers is less clear-cut, although a high-fat diet appears to increase the risk of both breast and prostatic cancers. It is very likely that some form of anti-hormone treatment will ultimately lessen the risk of these cancers.

Cancer is a multi-factorial disease and we are at the dawn of a major new understanding of the induction and development of the disease at the molecular level. The geneticists and molecular biologists will certainly be the first to understand these key events, and will devise new forms of biological intervention. They will surely also call upon the chemists and pharmacologists to design and produce chemical 'magic bullets' that will ultimately conquer the disease. In the meantime, the best advice would be: eat a healthy diet, avoid obvious carcinogens and viruses and use self-examination and the available screening facilities where appropriate.

Chapter 5

Magic Bullets: Still Elusive After All These Years

One hundred years after Paul Ehrlich introduced the idea of magic bullets, a convincing example of this species eludes us. Some of the beta-lactam antibiotics, and perhaps the anti-herpes drug acyclovir, come close to satisfying the criteria laid down by Ehrlich. Most success has been achieved with drugs, like these, that are specific inhibitors of key bacterial or viral enzymes. Even when this has been possible, the bacteria and viruses have usually responded by producing new or modified enzymes, and have thus become resistant to the drugs. We must be optimistic that the discovery of new antibacterial agents, like the recently described oxazolidinones, which inhibit protein biosynthesis at points not affected by other drugs, will be efficacious, at least for a while. In addition, as we gain more knowledge about the constitution of the bacterial cell wall and its assembly, it should be possible to design drugs that will further exploit the major differences that exist between these cells and our own.

In the longer term, success will surely follow from a greater understanding of the intimate details of how genes are 'switched on and off'. For example, it is already clear that bacteria usually express so-called virulence genes after invading a human cell, but do not necessarily express these genes when growing in culture. These virulence genes are obvious targets for drug attack. For cancer chemotherapy, the growing knowledge of the functions of the tumour suppressor genes and oncogenes should provide dividends in terms of the design of drugs that will modify their functions.

As mentioned in the last chapter, and illustrated in Fig. 4.5, when growth factors interact with receptors on the outer face of our cells, they trigger a cascade of chemical reactions that ultimately leads to activation of DNA transcription, and thence to production of m-RNA and new enzymes and proteins. Although this sequence of events is highly complex, a basic understanding of the key processes has been attained, and there are a number of drugs that interfere with these exquisitely complex signalling pathways. The human genome contains around 2000 genes that code for kinases and a further 300–500 that code for phosphatases; hence, it is not surprising that about one-third of all proteins and enzymes are regulated by reversible phosphorylation/dephosphorylation reactions catalysed by kinases/phosphatases. For anti-cancer therapy, the membrane-bound growth factor receptor tyrosine kinases have been prime targets for drug design since they are over-expressed in many cancers. Several main types of receptor kinase have been investigated and these include those associated with receptors for vascular endothelial growth factors (VEGF), epidermal growth factors (EGF), platelet derived growth factors (PDGF) and nerve growth factors (NGF). Each of these exists as a single polypeptide chain that spans the cell membrane. The extracellular portion carries the receptor and the cytoplasmic portion carries tyrosine residues that are autophosphorylated by the receptor's own tyrosine kinase.

Some natural products have specific inhibitory activity against these tyrosine kinases. For example, lavandustin from *Streptomyces griseolavendus* acts as a specific inhibitor of the EGF receptor kinase. Many of the synthetic compounds under development have structures that resemble adenosine

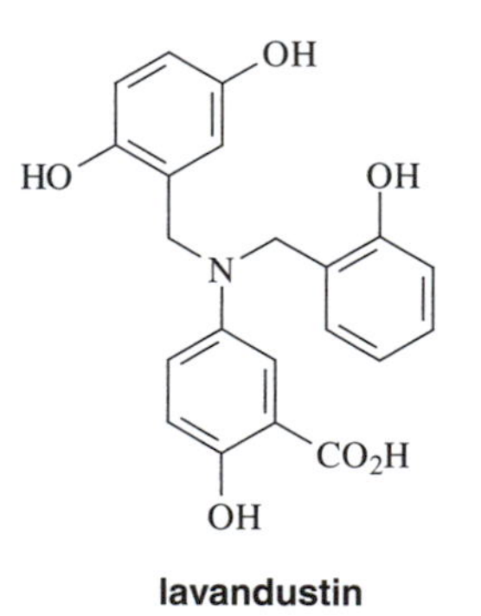

lavandustin

triphosphate (ATP) since this is the agent that most often carries out the phosphorylations catalysed by the kinases. Thus, the Astra Zeneca drug Iressa was the end product of a design programme that commenced with a screening of hundreds of AZ compounds that possessed a pyrimidine ring within their structures. The early lead compound *A* had an IC_{50} value of 5 nM against EGF receptor kinase, but was rapidly metabolised by hydroxylation at the benzylic position. The analogue *B* (which became Iressa) was more water-soluble and less readily metabolised although having a lower IC_{50} (23 nM), and it also had good selectivity for the EGF receptor kinase and for a related receptor called human epidermal growth factor receptor (HER-2), both of which can be over-expressed by as much as 100-fold in a number of cancers. At present, Iressa is licensed for use in the USA, Australia and Japan, and in this last country is the subject of a huge clinical trial involving more than 28,000 patients with lung cancer. A similar drug called Tarceva from Oncogene Sciences is currently in phase III clinical trials.

The other advanced drug candidate is Glivec from Novartis, which inhibits a membrane-bound kinase that is over-expressed in around 95% of patients with chronic myeloid leukaemia. As long ago as the 1960s, it was shown that these patients possessed a modified and shortened version of chromosome 22, which became known as the Philadelphia chromosome. In 1973, it was discovered that the missing piece of the chromosome had been relocated onto chromosome 9, and the protein product of this modified gene was an aberrant kinase, and this leads to continuous activation of white blood cell production. In 1990, work began at Ciba-Geigy to try to design general kinase inhibitors, which again contained a pyrimidine ring, and this research did identify at least one specific inhibitor of the CML kinase. Ciba-Geigy and Sandoz subse-

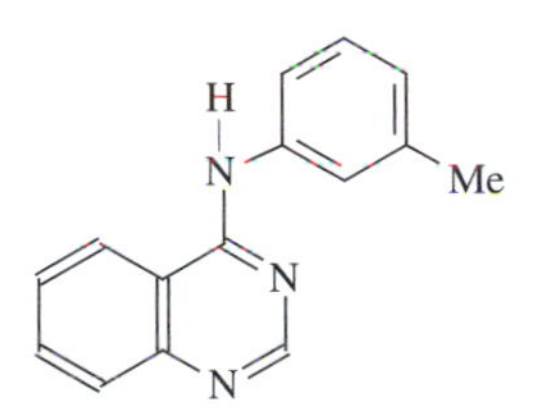

IC_{50} 0.18 micromolar

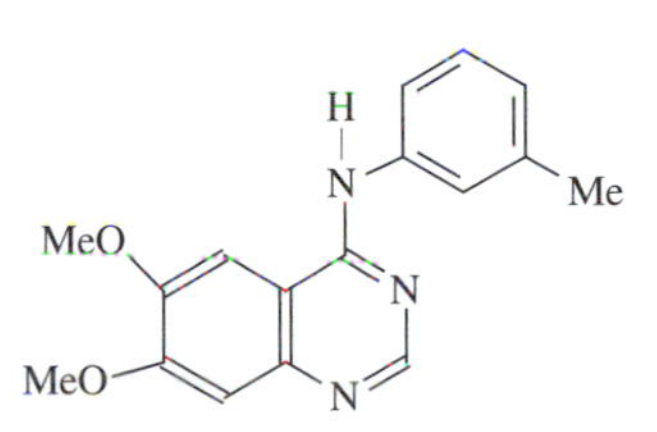

analogue A IC_{50} 5 nanomolar

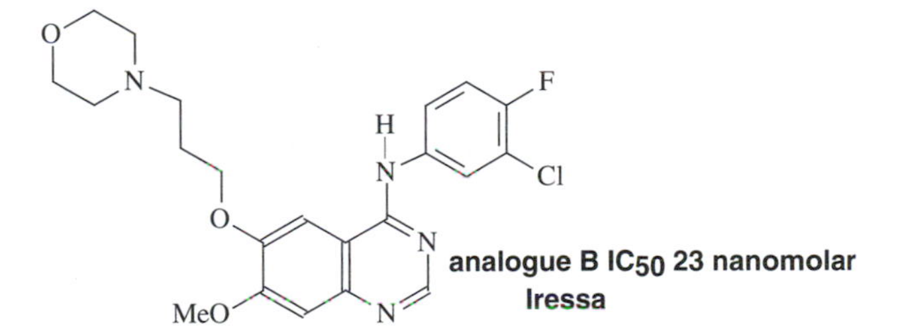

analogue B IC_{50} 23 nanomolar
Iressa

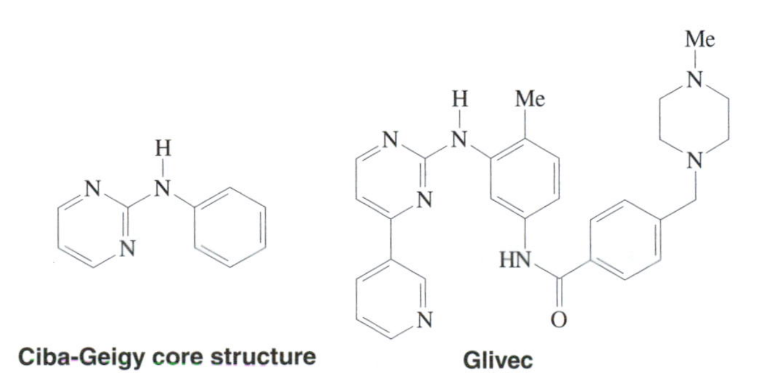

Ciba-Geigy core structure

Glivec

quently merged to form Novartis, and clinical trials of this drug (which became Glivec) began in the USA in 1998. The excellent clinical success of this new drug led to its approval for general CML patient use in the USA in December 2000 and in the UK in August 2002. Recent results suggest that Glivec given alone is about four times more effective than the older drug regime of γ-interferon and Ara-C in the reduction of white cell proliferation, and 75% fewer patients progress to late-stage disease. However, it is an expensive drug and costs about $28,000 per patient per year.

An excellent example of a family of drugs that inhibit a phosphatase rather than a kinase is provided by the immunosuppressant cyclosporins and FK506 (Tacrolimus). The former group of natural products, predominantly cyclosporins A and B, were isolated from the soil microorganisms *Cylindrocarpon lucidum* and *Tolypocladium inflatum* found in Wisconsin and

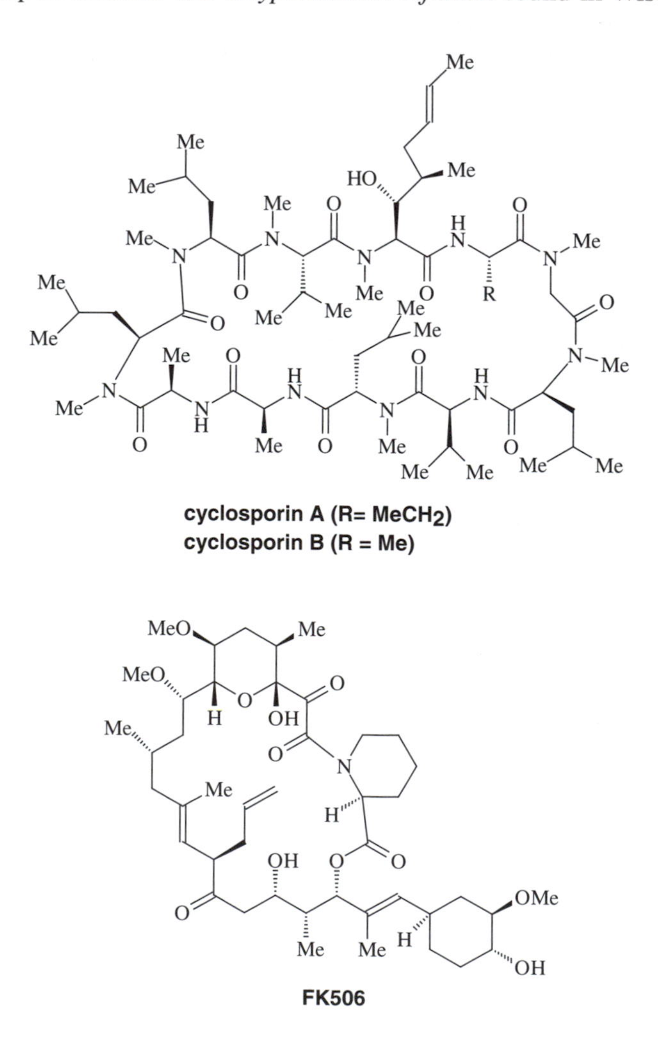

cyclosporin A (R= $MeCH_2$)
cyclosporin B (R = Me)

FK506

Norway, respectively, in 1970. FK506 was isolated from *Streptomyces tsukubaensis* (Japan) in 1984, and principally due to the research efforts of Stuart Schreiber and his group at Harvard, these compounds were shown to bind to the phosphatase calcineurin in the cytoplasm of T-lymphocytes (Fig. 5.1). Initially, they bind to small proteins to form complexes termed cyclosporin–cyclophylin and FK506 – FK binding protein, respectively, and

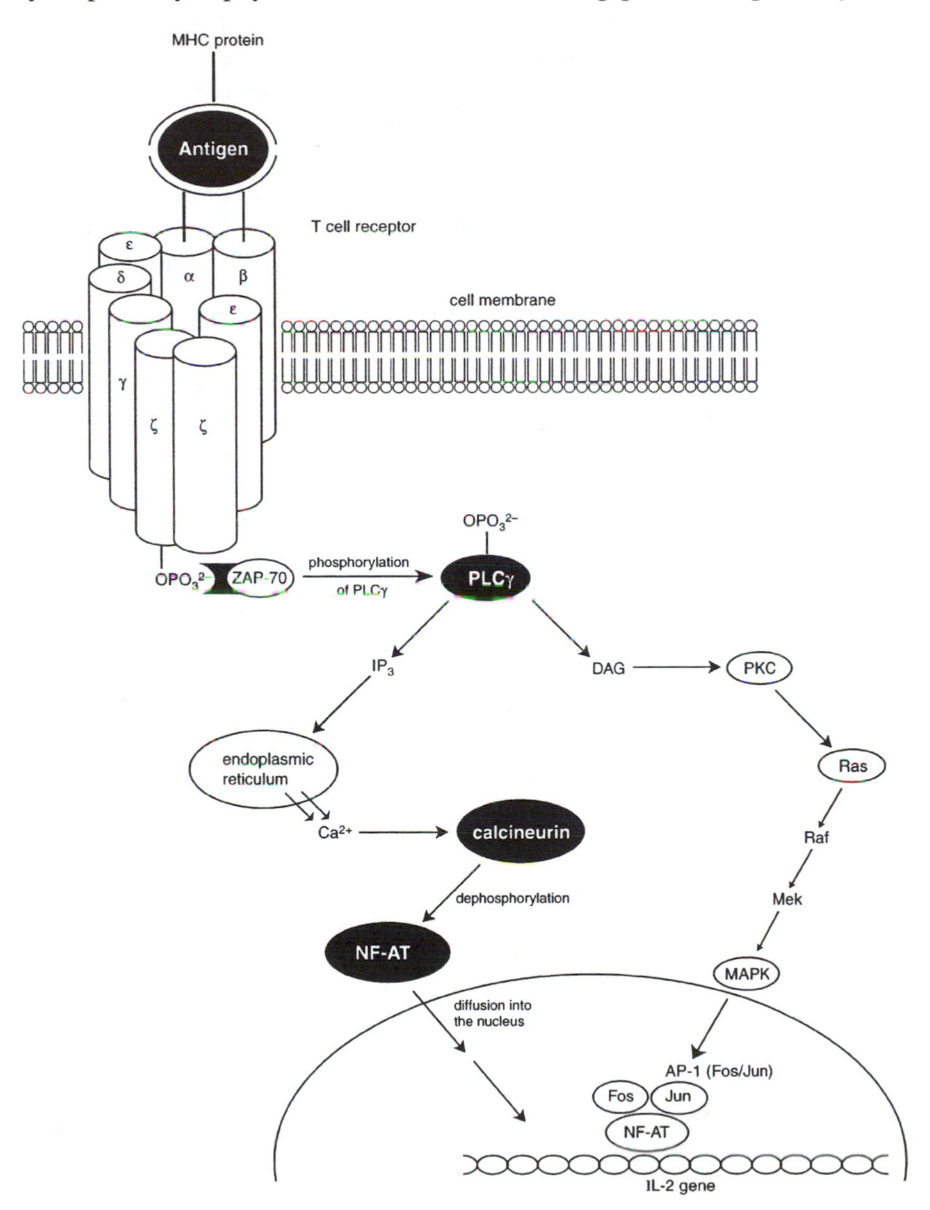

Figure 5.1

these then attach to calcineurin. Under normal circumstances, calcineurin is activated by calcium ions and then dephosphorylates (i.e., phosphatase activity) the transcription factor NFAT (nuclear factor of activated T-cells), which then passes into the nucleus of the T-cell and associates with the gene complex that codes for the protein interleukin-2 (Il-2). This gene product is responsible for the activation and reproduction of other T-lymphocyes and hence is important in the control of the immune response. Inhibition of calcineurin by cyclosporins and FK506 leads to a much reduced response and hence to their activities as immunosuppressants that have revolutionised the transplantation surgery.

While the targeting of kinases and phosphatases exemplifies our growing knowledge of the intricacies of cell signalling, it is potentially fraught with danger given the large number of these enzymes that are present in normal cells, and a new strategy involves inhibitions of a more select group of proteins. These are the so-called heat-shock proteins or molecular chaperones

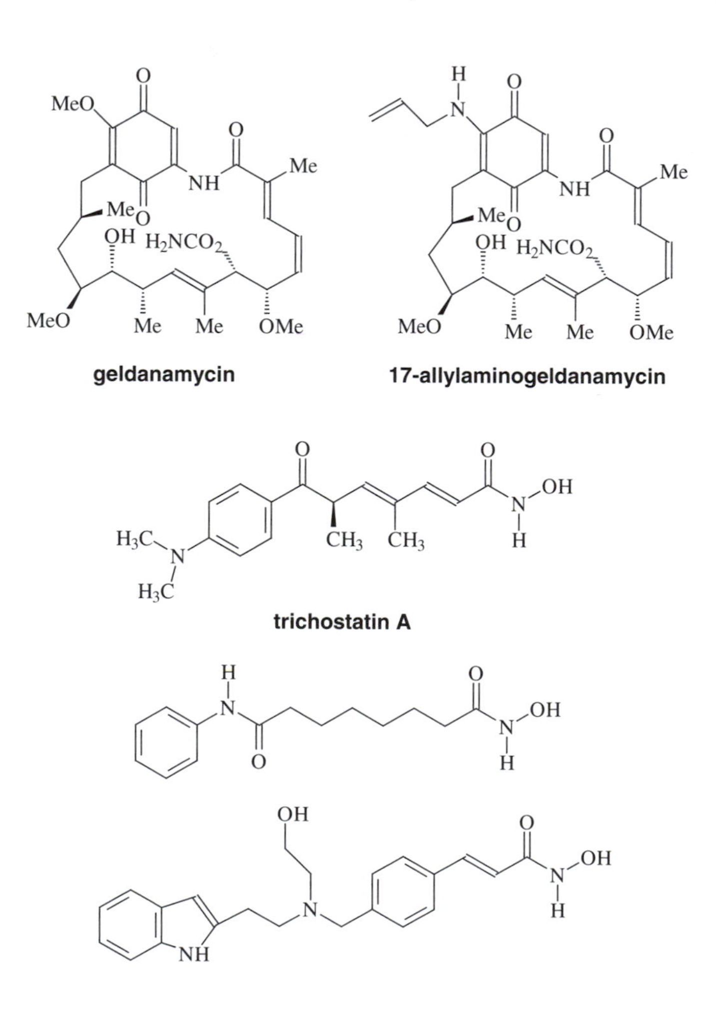

that protect cellular proteins from environmental and thermal damage, but that also assist newly formed proteins to attain their correct three-dimensional conformations. Many cancer cells produce large amounts of heat shock proteins (Hsps) and the search is on for molecules that will stop them from functioning normally. The molecular chaperone Hsp90 is especially good at protecting cells from damage from drugs and also controls various cell signalling pathways that ensure the cell's survival. In tumour cells, Hsp90 aggregates with other heat shock proteins to produce a super chaperone whilst this does not occur in normal cells, and at least one new drug – the 17-allylamino-analogue of the natural product geldanomycin – appears to undergo a conformational change (from an extended planar form to a more compact form) prior to binding to the super chaperone. A similar interaction with the non-aggregated form of Hsp90 does not occur, thus offering some prospect of specific activity in cancer cells.

An emerging class of targets for drugs are the histone deacetylases. Genes are wrapped around globular proteins called histones and these are decorated with acetyl and methyl groups, with phosphates, and even with ribose units. Some of these groups must be removed in order to change the conformation of the genes rendering them more accessible to transcription factors. The enzymes responsible for removing acetyl groups – the histone deacetylases – appear to be pre-eminent in the control of this process, and a number of inhibitors are showing considerable promise in the clinic. Prominent amongst these are the simple hydroxamates like trichostatin A from *Streptomyces hygroscopicus*, and analogues modelled upon this natural product.

Some progress in the design of 'genetic medicines', like *antisense agents* and *triplex-forming drugs*, has also been made in recent years. The former comprise short sequences of chemically modified DNA or RNA molecules, that bind in a complementary fashion to m-RNA and prevent it from acting as a blueprint for protein biosynthesis. The triplex-forming drugs comprise a short sequence of DNA, complementary to sequences found at the site of transcription on chromosomal DNA. These drugs bind to this DNA and induce the formation of triple helical sections, and this leads to disruption of transcription (see Fig. 5.2.). The drawbacks of these two classes of agents are their relative expense, their poor uptake by cells, and their ready destruction by DNAase enzymes. But they do demonstrate that it is possible to effect selective disruption of gene transcription or translation, and some success has been reported in clinical trials with antisense therapy for Crohn's disease, non-Hodgkin lymphoma and ovarian cancer.

More recently, it has been discovered that certain natural m-RNA sequences (termed *aptamers*) can inhibit their own ability to act as a blueprint, following binding of metal ions or small organic molecules. Clearly,

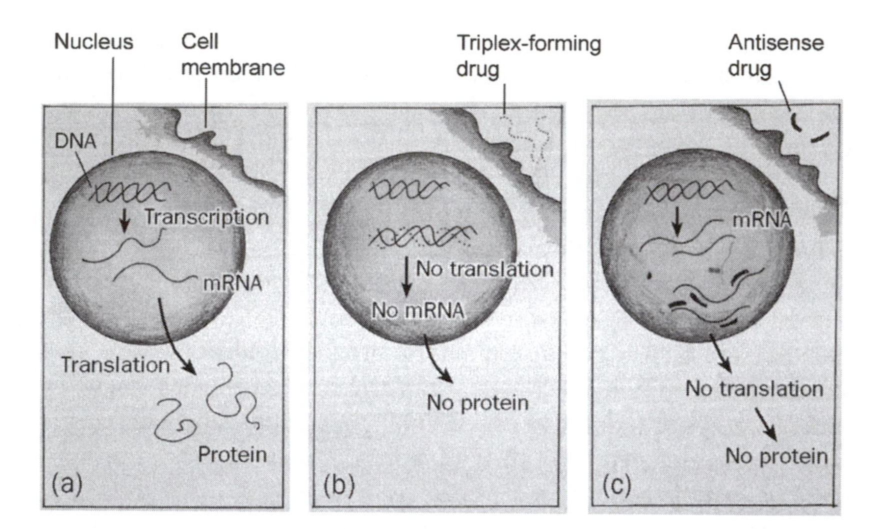

Figure 5.2 *Anti-sense and triplex-forming drugs. (a) Normal cell, (b) cell in the presence of triplex-forming drug, (c) cell in the presence of anti-sense drug*

the design of specific compounds that bind to these aptamers could be highly relevant for therapy.

Of course, it is also possible to transfer whole therapeutic or replacement genes into cells, and this has already been accomplished by using viruses like vaccinia to smuggle them in as part of its DNA complement. Non-viral technology includes the use of key DNA fragments encapsulated within cyclodextrin polymers or within liposomes. None of the viral vectors have ideal properties. Adenoviruses, for example, enter cells efficiently and produce a high expression of the therapeutic gene without becoming integrated into the host cell DNA. On the downside, they may induce an immune response. Retroviral vectors are also efficiently taken up by cells and their DNA integrates with host cell DNA, but they are hard to produce and undergo frequent random mutations. The recent discovery that p53 suppressor genes induce apoptosis, at least in part by coding for the production of reactive highly toxic radicals, suggests that it should be possible to smuggle in other genes that specifically code for the production of these species. But these strategies are, at present, difficult to use, except in the localised treatment of tumours.

The Human Genome Project has revealed that our genome contains between 30 and 40,000 genes – perhaps only 50% more than the genome of the fruit fly. This obviously requires that each gene codes for more than one protein or enzyme, and this is easily achieved if transcription can begin and end at different points along the gene. In addition, since much of the DNA sequence is untranscribed – sections known as *extrons* that are non-coding –

then the transcription process can splice together different coding regions (*introns*) whilst removing the extrons. The actual proteins and enzymes that result can be further modified through cycles of phosphorylation and dephosphorylation, but they can also be glycosylated, acetylated or attached to various lipids, thus increasing the number of different proteins that can be produced.

The challenge of the new science of *proteomics* will be to identify these novel products of transcription, especially when they are produced in aberrant cells associated with conditions like cancer and arthritis. Since it has been estimated that as many as 10 genes may influence the course of a disease, there is clearly much work for the chemical biologist and synthetic chemist. They will have three clear objectives:

- Identification of lead structures that will modify gene action or the processes by which they are transcribed and translated.
- Optimisation of these structures until clinical candidates are identified.
- Design of synthetic routes that allow large quantities of these compounds to be prepared cheaply and efficiently.

Despite the present apparent unpopularity of chemistry in schools and colleges, it is difficult to see how any of these objectives can be realised by anyone other than highly skilled chemists. Their research will be as exciting and important as that which preceded it on the Human Genome Project.

On a more mundane yet increasingly relevant level, our growing awareness of the importance of environmental factors will also be crucial for the treatment of disease, especially emerging diseases. It is, for example, inconceivable that the appearance today of Legionnaires' disease, would cause as much confusion as it did in July 1976. We are much more *au fait* with the ability of microorganisms (*Legionella pneumophila* in this instance) to exploit man-made ecological niches (like air conditioning cooling towers). The recognition of the underlying cause of a disease is thus crucial for the design of new chemotherapeutic agents and treatment strategies

One recent example of a specific drug therapy that evolved, once the underlying cause of the disease had been established, is provided by the inventions of therapies to treat peptic ulcer disease. Around 10% of the population will develop a chronic peptic ulcer at some stage in their lives, and as well as causing severe discomfort and pain, the condition may progress to stomach cancer. A major triumph of rational drug design, in the early 1970s, was the invention of the drug cimetidine (Tagamet), by the group of Robin Ganellin, John Emmett and Graham Durant at Smith-Kline and French, in Welwyn Garden City. This is a specific antagonist of the histamine (H2) receptors in the gut, and leads to a reduction in the release of gastric acid.

Cimetidine, and other similar drugs like Glaxo's ranitidine (Zantac), have revolutionised the treatment of peptic ulcer disease, as well as making fortunes for the companies. The drugs, however, treat the disease without tackling the underlying cause.

In 1982, Barry Marshall, a physician at the Queen Elizabeth II Hospital in Perth, Western Australia, first isolated the bacterium *Campylobacter pylori* (now *Helicobacter pylori*) from the stomachs of patients suffering from severe gastritis or peptic ulcer disease. In the subsequent years, it has been demonstrated that around 50% of the world's population (perhaps 80%, in the developing countries) harbour this bacterium. The bacterium probably causes disease through its production of enzymes and toxins, but also through the inflammatory products of the immune response that it induces.

Once this bacterial origin had been established, alternative therapies could be attempted. These days, complete eradication of the bacterium, with a resultant cure of disease, can be effected through the use of a mixture of anti-bacterial agents, like tetracycline and metronidazole, in conjunction with finely divided bismuth citrate, to which the bacterium is exquisitely sensitive. Whilst this drug cocktail would probably not qualify for the title 'magic bullet', Paul Ehrlich and Paracelsus would surely approve of the use of a salt of the metal bismuth!

When we review the advances made in chemotherapy during the past century, one is left in awe by what has been achieved. In 1900, there were only a handful of pure drugs with proven efficacy: quinine, morphine, heroin, cocaine and aspirin. Today, there are literally thousands of drugs with good levels of efficacy. The life-threatening potential of bacterial infections has, to a large extent, been overcome, although the viruses and cancer still pose major threats. We are now very close to understanding how disease arises and progresses at the genetic level. We should then be well-placed to make manipulations at the molecular level, and design real 'magic bullets'.

Francis Bacon (1561–1626) foresaw the advantages of a complete understanding of disease causation four centuries ago:

> As if you would call a physician, that is thought good for the cure of the disease you complain of but is unacquainted with your body...and so cure the disease and kill the patient.

During this new century, we shall surely obtain a full understanding, at the genetic level, of how the body functions in both health and disease, and then discover how to cure the disease *and* save the patient.

Further Reading

For most topics, major and readily accessible books and journals are cited, and these will facilitate access to more detailed sources of information. The entries are listed in the order that the subject matter is covered in each chapter. For the latest research references the reader is referred to internet sources like rsc.org, pubs.acs.org. or elsevier.com, but the most useful review articles have been included.

Chapter 1

Ball, P. (2001). The life of Paracelsus. *Chem. Brit.*, **37**, 40.

Colebrook, L. and Fleming, A. (1911). A clinical trial of Salvarsan. *Lancet*, **i**, 1631.

Dale, H., Browning, C. H., and Marquardt, M. (1954). Lectures delivered to mark the centenary of Ehrlich's birth. *Lancet*, **i**, 4863.

Ehrlich, P. (1913). A lecture on chemotherapeutics. *Lancet*, **ii**, 445.

Hardman, J. G. and Limbard, L. E. (1996). *Goodman and Gilman's the pharmacological basis of therapeutics*, 9th edn. McGraw-Hill, New York.

Mann, J. (2000). *Murder, magic and medicine*, 2nd edn. Oxford University Press, Oxford.

Marquardt, M. (1949). *Paul Ehrlich*. Heinemann, London.

Parascandola, J. (1981). The theoretical basis of Ehrlich's chemotherapy. *J. Hist. Med.*, **36**, 19.

Patrick, G. L. (2001). *An introduction to medicinal chemistry*, 2nd edn, Oxford University Press, Oxford.

Sneader, W. (1996). *Drug prototypes and their exploitation*. Wiley, Chichester.

Thomas, G. (2000). *Medicinal Chemistry*. Wiley, Chichester.

Ward, P. S. (1981). The American reception of Salvarsan. *J. Hist. Med.*, **36**, 44.

Weatherall, M. (1990). *In search of a cure: A history of pharmaceutical discovery*. Oxford University Press, Oxford.

Chapter 2

Abraham E. P. and Newton, G. G. F. (1961). *Biochem. J.*, Structure elucidation of cephalosporin C. **79**, 377.

Aldridge, S. (2000). The discovery of penicillin. *Chem. Brit.*, **36**, 32.

Bachelor, F. R., Doyle, F. P., Nayler, J. H. C., and Rolinson, G. N. (1959). Discovery of 6-APA in penicillin fermentation. *Nature*, **183**, 257.

Bentley P. H. and O'Hanlon P. J. (eds), (1997). *Anti-infectives: recent advances in chemistry and structure–activity relationships*. RSC, Cambridge.

Blessington B. and O'Sullivan, J. O. (1994). The captain of death returns. *Chem. Brit.*, **30**, 566.

Brickner, S. J. (1997). Multi-resistant bacterial infections — driving the search for new antibiotics. *Chem. Ind.*, 131.

Brock, T. D. (1988). *Robert Koch*. Springer, Berlin.

Brown, T. M., Dronsfield A. T., and Ellis, P. M. (2004). Conquering TB, the 'captain of death'. *Educ. Chem.*, 15.

Chauvette, R. R., Pennington, P. A., Ryan, C. W., Cooper, R. D. G., Jose, F. L., Wright, I. G., van Heyningen, E. M., and Huffman, G. W. (1971). Conversion of penicillins into cephalexin. *J. Org. Chem.*, **36**, 1259.

Christensen, B. G., (1989). Tienam (thienamycin) — from natural product to antibiotic. *Chem. Brit.*, **25**, 371.

Cimarusti C. M. and Sykes, R. B. (1983). Monobactams — novel antibiotics. *Chem. Brit.*, **19**, 302.

Cohen, M. L. (1992). Epidemiology of drug resistance. *Science*, **257**, 1050.

Dubos, R. (1960). *Louis Pasteur: freelance of science*. da Capo Press, New York.

Duncan, K. (1997). Towards the next generation of drugs and vaccines for tuberculosis. *Chem. Ind.*, 861.

Durckheimer, W., Blumbach, J., Lattrell, R., and Scheunemann, K. H. (1985). *Ang. Chem. Int. Edn.*, Recent developments in the field of beta-lactam antibiotics. **24**, 180.

Evans, J. (1998). TB — know your enemy. *Chem. Brit.*, **34**, 38.

Finch, R. R., Hill, P., and Williams, P. (1995). Staphylococci — the emerging threat. *Chem. Ind.*, 225.

Fleming, A. (ed) (1950). *Penicillin*. Butterworth, London.

Florey, H. W. (1945). Use of microorganisms for therapeutic purposes. *BMJ*, **ii**, 4427.

Ford, C. (2001). First of a kind — linezolid. *Chem. Brit.*, **38**, 22.

Garrett, L. (1994). *The coming plague*. Penguin Books, New York.

Gilpin, M. L. and Milner, P. H. (1997). Resisting change — the glycopeptide antibiotics. *Chem. Brit.*, **33**, 46.

Goodlee, R. J. (1924). *Lord Lister*. Oxford University Press, Oxford.

Grohe, K. (1992). Antibiotics — the new generation (fluoroquinolones). *Chem. Brit.*, **28**, 34.

Guthrie, D. (1949). *Lord Lister*. Livingstone, Edinburgh.

Hayes, J. D. and Wolf, C. R. (1990). Mechanisms of drug resistance. *Biochem. J.*, **272**, 281.

Heatley, N., Corbett, K., and Bentley, R. (1990). A review of the work on penicillin carried out at Oxford. *The Biochemist*, **12**, 4.

Henry, C. M. (2000). Antibiotic resistance. *Chem. Eng. News*, March 6, 41.

Hook, V. (1997). Superbugs step up the pace. *Chem. Brit.*, **33**, 34.

Huebner, R. E. and Castro, K. G. (1995). The changing face of tuberculosis. *Ann. Rev. Med.*, **46**, 47.

MacFarlane, R. G. (1979). *Howard Florey: the making of a great scientist*. Oxford University Press, Oxford.

MacFarlane, R. G. (1985). *Alexander Fleming: the man and the myth*. Oxford University Press, Oxford.

Mann, J. (1995). Rhe antibiotic era. *Educ. Chem.*, 94.

Mann, J. and Crabbe, M. J. C. (1996). *Bacteria and antibacterial agents*. Oxford University Press, Oxford.

Massova, I. and Mobasherry, S. (1997), Molecular basis of interactions between beta-lactam antibiotics and beta-lactamases. *Acc. Chem. Res.*, **30**, 162.

Newall, C. E. (1987). Ceftazidime – an injectable antibiotic. *Chem. Brit.*, **23**, 976.

Niccolai, D., Tarsi, L., and Thomas, R. J. (1997). The renewed challenge of antibacterial chemotherapy. *Chem. Commun.*, 2333.

Quirke, V. (1998). Howard Florey — medicine maker. *Chem. Brit.*, **34**, 35.
Rouhi, A. M. (1999). Tuberculosis — a tough adversary. *Chem. Eng, News*, May 17, 52.
Sheehan, R. (1982). *The enchanted ring; the untold story of penicillin*. MIT Press, Cambridge, MA.
Shrewsbury, J. F. D. (1970). *A history of the Bubonic Plague in the British Isles*. Cambridge University Press, Cambridge.
Staunton, J. and Weissman, K. J. (2001). Polyketide biosynthesis. *Nat. Prod. Rep.*, **18**, 380.
Stinson, S. C. (1996). Drugs firms restock antibacterial arsenal. *Chem. Eng. News*, September 23, 75.
(various authors) (1994). Revising the antibiotic miracle. *Science*, **264**, 360.
(various authors) (1999). Microbes, immunity and disease. *Science*, **284**, 1302.
Walsh, C. (1999). Deconstructing vancomycin. *Nature*, **284**, 442.
Williams, D. H. (1996). The glycopeptide story — how to kill the deadly superbugs. *Nat. Prod. Rep*. **13**, 469.
Williams, D. H. (1999). Sugaring vancomycin. *Nature*, **397**, 567.
Wright, G. D. (2003). Mechanisms of resistance to antibiotics. *Curr. Opin. Chem. Biol.*, **7**, 563.
Young, D. B. (1998). *Nature*, Blueprint for the white plague. **393**, 515.
Yuan, Z. (2000). Actinonin — a deformylase inhibitor. *Biochemistry*, **39**, 1256.

Chapter 3

Anderson, R. (1993). AIDS — trends, predictions, controversy. *Nature*, **363**, 393.
Antia, R., Regoes, R. R., Koella, J. C., and Beergstrom, C. T. (2003). The role of evolution in the emmergence of infectious diseases. *Nature*, **426**, 658.
Armelagos, G. J. (1998). The viral superhighway. *The Sciences*, Jan/Feb, 24.
Ashton, V. (2003). Cold comfort. *Chem. Brit.*, **39**, 26.
Bella, J. and Rossmann, M. G. (1998). Rhinoviruses and their ICAM receptors. *Pharm. News*, **5**, 22.
Berger, E. A., Doms, R. W., Fenyo, E.-M., Korber, B. T. M., Littman, D. R., Moore, J. P., Sattenau, Q. J., Schuitemarker, H., Sodroski, J., and Weiss, R. A. (1998). A new classification for HIV-1. *Nature*, **391**, 240.
Beveridge, W. I. B. (1977). *Influenza: the last great plague*. Heinemann, London.
Challand, R. (1997). Antiviral chemotherapy. *Educ. Chem.*, 99.
Challand, R. and Young, R. J. (1997). *Antiviral chemotherapy*. Spektrum, Oxford.
Clapham, P. R. and Weiss, R. A. (1997). Spoilt for choice of co-receptors. *Nature*, **388**, 230.
Darby, S. C., Ewart, D. W., Giangrande, P. L. F., Dolin, P. J., Spooner, R. J. D., and Rizza, C. R. (1995). Mortality of haemophiliacs infected with HIV. *Nature*, **377**, 79.
de Cl ercq, E. (2002). New developments in anti-HIV chemotherapy. *Biochim. Biophys. Acta.*, **1587**, 258.
de Clercq, E. (2004). Antiviral drugs in clinical use. *J. Clin. Virol.*, **30**, 115.
Duesberg, P. (1991). AIDS epidemiology — inconsistencies with HIV and with infectious disease. *Proc. Natl. Acad. Sci. USA*, **88**, 1575.
(Editorial) (1983). No need to panic about AIDS. *Nature*, **302**, 749.
(Editorial) (1983). Human T-cell leukaemia virus linked to AIDS. *Science*, **220**, 806.
(Editorial) (1995). Duesberg and the new view of HIV. *Nature*, **373**, 189.
(Editorial) (1996). New hope in HIV disease. *Science*, **274**, 1988.
(Editorial) (2003). SARS — what have we learned? *Nature*, **424**, 114.
Emon, T. D., Evans, B., Bowen, E. T. W., and Lloyd, G. (1977). A case of Ebola virus infection. *BMJ*, **ii**, 541.

Fauci, A. S. (1996). Host factors and the pathogenesis of HIV-induced disease. *Nature*, **384**, 529.

Fischer, R. B. (1991). *Edward Jenner*. Andrew Deutsche, New York.

Gibaldi, M. (1996). Progress in the treatment of HIV infection. *Pharm. News*, **3**, 27.

Gibaldi, M. (1997). The search for markers of HIV disease. *Pharm. News*, **4**, 24.

Gopinath, L. (1997). Drug cocktails to fight HIV. *Chem. Brit.*, **33**, 38.

Groopman, J. E. and Gottlieb, M. S. (1983). AIDS – the widening gyre. *Nature*, **303**, 575.

Guan, Y. *et al.* (2004). H5N1 influenza — a protean pandemic threat. *Proc. Nat. Acad. Sci.*, **101**, 8156.

Le. Guenno, B. (1995). Emerging viruses. *Sci. Amer.*, **273**, Oct., 30.

Harnden, M. R. (ed) (1985). *Approaches to antiviral agents*. MacMillan, Basingstoke.

Hopkins, D. R. (1988). *Princes and peasants*. University of Chicago Press, Chicago.

Johnson, H. M., Bazer, F. W., Scente, B. E., and Jarpe, M. A. (1994). How interferons fight disease. *Sci. Amer.*, **270**, May, 40.

Kilbourne, E. D. (1987). *Influenza*. Plenum, New York.

Laver, W. G., Bischofberger, N., and Webster, R. G. (1999). Disarming flu viruses. *Sci. Amer.*, **280**, 56.

Levine, M. M. (1996). The legacy of Edward Jenner. *Lancet*, **312**, 1177.

Matthews, T., Salgo, M., Greenberg, M., Chung, J. deMasi, R., and Bolognesis, D. (2004). Enfuvirtide — the first therapy to inhibit the entry of HIV-1 into host CD4 lymphocytes. *Nat. Rev. Drug Disc.*, **3**, 215.

McClure, M. (1989). AIDS and the monkey puzzle. *New Sci.*, 25 March, 46.

Miller, G. (1957). *The adoption of inoculation for smallpox in England and France*. Oxford University Press, Oxford.

Mitsuya, H., Yarchoan, R., and Broder, S. (1990). Molecular targets for AIDS therapy. *Science*, **249**, 1533.

Monath, T. P., Newhouse, V. F., Kemp, G. E., Setzer, H. W., and Cacciapuoti, A., (1974). Lassa virus isolation from *Mastomys natalensis* rodents during an epidemic in Sierra Leone. *Science*, **185**, 262.

Monto, A. S. (2003). The role of antivirals in the control of influenza. *Vaccine*, **21**, 1796.

Nabel, G. J. (1999), Surviving Ebola virus infection. *Nat. Med.*, **5**, 373.

Nelson, S., (1991). AIDS research — travelling hopefully. *Chem. Brit.*, **27**, 294.

Nicholson, B. (ed) (1994). *Synthetic vaccines*. Blackwell, Oxford.

Nowak, M. A. and McMichael, A. J. (1995). How HIV defeats the immune system. *Sci. Amer.*, **273**, 42.

Oxford, J. S., Sefton, A., Jackson, R., Innes, W., Daniels, R. S., and Johnson, N. P. A. S. (2002). World War I may have allowed the emergence of Spanish influenza. *Lancet Infect. Dis.*, **2**, 111.

Penn, M. L., Grivel, J-C., Schramm, B., Goldsmith, M. A., and Margolis, L., (1999). Utilisation of CXCR4 in HIV-1 infected human lymphoid tissue. *Proc. Natl. Acad. Sci. USA*, **96**, 663.

Pestka, S. (1983). The purification and manufacture of human interferons. *Sci. Amer.*, **49**, August, 29.

Rabinovich, N. R., McInnes, P., Klein, D. L., and Hall, B. F. (1994). Vaccine technologies — view to the future. *Science*, **265**, 1401.

Reid, A. H., Taubernberger, J. K., and Fanning, T. G. (2001). *Microbes Infection*, **3**, 81.

Richman, D. D. (1996). HIV therapeutics. *Science*, **272**, 1886.

Rosenberg, Z. and Fauci, A. (1990). Inside the AIDS virus. *New Sci.*, 10 Feb., 51.

Skehel, J. J. *et al.*, (2004). Structure and receptor-binding of 1918 flu haemagglutinin. *Science*, **303**, 1838.

Stevens, J., Corper, A. L., Basler, C. F., Taubenberger, J. K., Palese, P., and Wilson, I. A. (2004). Structure of haemagglutinin from the 1918 influenza virus. *Science*, **303**, 1866.

(Various authors) (1998). Special update on AIDS research. *Science*, **280**, 1855.

von Itstein, M. (1993). *et al.*, Rational design of neuraminidase inhibitors. *Nature*, **363**, 418.

Wain-Hobson, S. (1998). 1959 and all that. *Nature*, **391**, 531.

Walker, E. and Christie, P. (1998). Chinese avian influenza. *BMJ*, **316**, 325.

Webby, R. J. and Webster, R. G. (2003). Are we ready for pandemic influenza? *Science*, **302**, 1519.

Webster, R. G. (1999). 1918 Spanish influenza — the secrets remain elusive. *Proc. Natl. Acad. Sci. USA*, **96**, 1164.

Webster, R. G. (2002). The importance of animal influenza for human disease. *Vaccine*, **20**, S16.

Weiss, R. A. and Jaffe, H. W. (1990). Duesberg, HIV and AIDS. *Nature*, **345**, 659.

Weiss, R. A. (1993). How does HIV cause AIDS? *Science*, **260**, 1273.

Zhu, T., Korbert, B. T., Nahmias, A. J., Hoper, E., Shapr, P. M., and Ho, D. D. (1998). An African HIV-1 sequence from 1959 and implications for the origin of the epidemic. *Nature*, **391**, 594.

Chapters 4 and 5

Abdulla, S. (1997). Silencing the code (antisense drugs). *Chem. Brit.*, **33**, 30.

Altmann, K.-H. (2004). Chemistry and chemical biology of the epothilones. *Org. Biomol. Chem.*, **2**, in press.

Ames, B. N., Gold, L. S. D., and Willett, W. C. (1995). The causes and prevention of cancer. *Proc. Natl. Acad. Sci. USA*, **92**, 5258.

Barnard, C. F. J., Cleare, M. J., and Hydes, P. C. (1986). Second generation anticancer platinum compounds. *Chem. Brit.*, **22**, 1001.

Beatson, G. T. (1896). The treatment of inoperable cases of carcinoma of the breast. *Lancet*, **ii**, 104, 162.

Bell, I. M. (2004). Inhibitors of farnesyl transferase. *J. Med. Chem.*, **47**, 1869.

Bergman, R. G. (1973). The Bergman cyclisation. *Acc. Chem. Res.*, **6**, 25.

Blattman, J. N. and Greenberg, P. D. (2004), Cancer immunotherapy. *Science*, **305**, 200.

Bonner, J. (2001). Cancer vaccines. *Chem. Ind.*, 277.

Boyle, F. T. and Costello, G. F. (1998). *Chem. Soc. Rev.*, **27**, 251.

Brazidec, J-Y. *et al.* (2004). Synthesis and biological evaluation of new geldanamycin derivative as potent inhibitors of Hsp90. *J. Med. Chem.*, **47**, 3865.

Bross, P. F. *et al.*, (2001). Approval summary for Mylotarg. *Clin. Cancer Res.*, **7**, 1490.

Burger, R. M. (1998). Cleavage of nucleic acids by bleomycin. *Chem. Rev.*, **98**, 1153.

Burke, M. (2003). On delivery (of genes). *Chem. Brit.*, **39**, 36.

Burnett, A. K. and Eden, O. B. (1997). The treatment of acute leukaemia. *Lancet*, **i**, 270.

Cassady, J. M. and Douros, J. D. (ed) (1980). *Anticancer agents based on natural product models*. Academic Press, New York.

Cech, T. R. (2004). RNA finds a simpler way (aptamers). *Nature*, **428**, 263.

Chen, A. Y. and Lui, L. F. (1994). DNA topoisomerases. *Ann. Rev. Pharm. Toxic.*, **34**, 191.

Chene, P. (2003). Inhibiting the p53-MDM2 interaction. *Nature Rev. Cancer*, **3**, 102.

Cirla, A. and Mann, J. (2003). Combretastatins. *Nat. Prod. Rept.*, **20**, 558.

Cohen, J. S. and Hogan, M. E. (1994). The new genetic medicines. *Sci. Amer.*, **271**, 51.

Cyranoski, D. (2003), Bespoke cancer drugs (Iressa and Tarceva). *Nature*, **423**, 209.

Dacre, J. C. and Goldmann, M. (1996). Toxicology and pharmacology of sulphur mustards. *Pharmacol. Rev.*, **48,** 289.

Denissenko, M. F., Pao, A., Tang, M.-S., and Pfeifer, G. P. (1996). Preferential formation of benzo[*a*]pyrene adducts at lung cancer mutational hotspots in *p53*. *Science*, **272**, 430.

Dodd, E. C., Goldberg, L., Lawson, W., and Robinson, R. (1938). Estrogenic activity of stilbenes. *Nature*, **141**, 247.

Doll, R. and Peto, R. (1981). *The causes of cancer.* Oxford University Press, Oxford.

Downward, J. (2003). Targeting ras signalling pathways in cancer chemotherapy. *Nat. Rev. Cancer*, **3**, 11.

Early Breast Cancer Trialist's Collaborative Group, (1992). Systematic treatment of early breast cancer by hormonal, cytotoxic, or immune therapy. *Lancet*, **i**, 1, 73

(Editorial), (1997). The prostate question, unanswered still. *Lancet*, **i**, 443

Evans, J. (2001). Where do we go from here (with new genetic medicines)? *Chem. Brit.*, **37**, 26.

Eyster, K. M. (1998). Introduction to signal transduction. *Biochem. Pharmacol.*, **55**, 1927.

Faulkner, D. J. (2000). Highlights of marine natural products chemistry 1972–1999. *Nat. Prod. Rept.*, **17**, 1.

Folkman, J. and Kalluri, R. Endogenous angiogenesis inhibitors. *Nature*, **427**, 787.

Garnick, M. B. (1994). The dilemmas of prostate cancer. *Sci. Amer.*, **267**, 52.

Gilman, A. and Phillips, F. S. (1946). Use of N-mustards. *Science*, **103**, 409.

Goodman, J. and Walsh, V. (2001). *The story of taxol.* Cambridge University Press, Cammbridge.

Griffiths, J. (2004). Photodynamic therapy. *Educ. Chem.*, May,71.

Haag, R. (2004). Supramolecular (tumour-specific) drug delivery. *Angew. Chem. Int. Edn.*, **43**, 278.

Haber, L. F. (1986). *The poisonous cloud: chemical warfare in the First World War*. Oxford University Press, Oxford.

Hackshaw, A. K., Law, M. R., and Wald, N. J. (1997). The accumulated evidence on lung cancer and environmental tobacco smoke. *BMJ*, **315**, 980.

Haddow, A. (1947). Chemistry of carcinogenic compounds. *BMJ*, **4**, 417.

Hale, K. J., Hummersone, M. G., Manaviazar, S., and Frigerio, M. (2002). Chemistry and biology of the bryostatins. *Nat. Prod. Rept.*, **19**, 413.

Harper, M. J. K. and Walpole, A. L. (1966). Contrasting endocrine activities of trisubstituted triphenylethylenes (tamoxifen progenitors). *Nature*, **212**, 87.

Hashimoto, S. and Hecht, S. M. (2001). Kinetics of DNA cleavage by Fe(II)-bleomycins. *J. Amer. Chem. Soc.*, **123**, 7437.

Hausen, H. Z. (1991). Viruses in human cancer. *Science*, **254**, 1167.

Huggins, C. and Hodges, C. V. (1941). Studies on prostate cancer. *Cancer Res.*, **1**, 293.

Jarman, M., Smith, H. J., Nicholls, P. J., and Simons, C. (1998). Inhibitors of enzymes of androgen biosynthesis. *Nat. Prod. Rept.*, **15**, 495.

Johnson, I. S., Armstrong, J. G., Gorman, M., and Burnett, J. P. (1963). The vinca alkaloids — a new class of oncolytic agents. *Cancer Res.*, **23**, 1390.

Jordan, V. C. (2003). The new antiestrogens and estrogen receptor modulators. *J. Med. Chem.*, **46**, 1081.

Kamal, A. *et al.*, (2003). Binding of geldanamycin to Hsp90. *Nature*, **425**, 407.

Kingston, D. G. I. (2001). Taxol, a molecule for all seasons. *Chem. Commun.*, 867.

LaFond, R. E. (ed), (1988). *Cancer: the outlaw cell*, 2nd edn. American Chemical Society, Washington.

Leonard, D. M. (1997). Ras-farnesyltransferase — a new therapeutic target. *J. Med. Chem.*, **40**, 2971.

Lerner, L. J. and Jordan, V. C. (1990). Development of antiestrogens and their use in breast cancer. *Cancer Res.*, **50**, 4177.

Levitzki, A. (2003). EGF receptor as a therapeutic target. *Lung Cancer*, **41**, S9.

Liekens, S., de Clercq, E., and Neyts, J. (2001). Angiogenesis — regulators and clinical applications. *Biochem. Pharmacol.*, **61**, 253.

Lown, J. W. (1993). Discovery and development of anthracycline antitumour antibiotics. *Chem. Soc. Rev.*, **22**, 165.

Macdonald, F. and Ford, C. H. J. (1997). *Molecular biology of cancer*. Bios, Oxford.

Mann, J. (2001). Natural products as immunosuppressive agents. *Nat. Prod. Rept.*, **18**, 417.

Mann, J. (2002), Natural products in cancer chemotherapy—past, present and future. *Nature Rev. Cancer*, **2**, 143.

Mapp, A. K. (2003). Regulating transcription — a chemical perspective. *Org. Biomol. Chem.*, **1**, 2217.

Marx, J. (1994). Oncogenes reach a milestone. *Science*, **266**, 1942.

Melton, L. (2004). Histone deacetylase inhibitors. *Chem. World*, April, 24.

Milgrom, L. and MacRobert, S. (1998). Photodynamic therapy. *Chem. Brit.*, **34**, 45.

Miller, T. A., Witter, D. J., and Belvedere, S. (2003). Histone deacetylase inhibitors. *J. Med. Chem.*, **46**, 5097.

Mu, F., Coffling, S. L., Riese, D. J., Geahlen, R. L., Verdier-Pinard, P., Hamel, E., Johnson, J., and Cushman, M. (2001). Design synthesis and biological evaluation of lavendustin A analogues. *J. Med. Chem.*, **44**, 441.

Mutter, R. and Wills, M. (2000). Chemistry and clinical biology of the bryostatins. *Bioorg. Med. Chem.*, **8**, 1841.

Neidle, S. and Waring, M. (eds) (1983). *Molecular aspects of anticancer drug action*. MacMillan, Basingstoke.

Newman, D. J., Cragg, G. M. and Snader, K. M. (2000). The influence of natural products upon drug discovery. *Nat. Prod. Rept.*, **17**, 215.

Nicolaou, K. C. and Dai, W.-M. (1991). Calicheamicin and other ene-diyne antibiotics. *Angew. Chem. Int. Edn.*, **30**, 1387.

Nicolaou, K. C., Roschangar, F., and Vourloumis, D. (1998). Chemical biology of epothilones. *Angew. Chemie. Int. Edn.*, **37**, 2014.

Nicolaou, K. C., Ritzen, A. and Namoto, K. (2001). Recent developments in the chemistry, biology and medicine of the epothilones. *Chem. Commun.*, 1523.

Niculescu-Duvaz, I. and Springer, C. J. (1997). ADEPT technology. *Adv. Drug. Delivery Rev.*, **26**, 151.

Novina, C. D. and Sharp, P. A. (2004). The RNAi revolution. *Nature*, **430**, 161.

Old, L. J. (1985). Tumour necrosis factor. *Science*, **230**, 630.

Olson, J. S. (1989). *The history of cancer: and annotated bibliography*. Greenwood Press, Westport.

Pettit, G. R., Pierson, F. H., and Herald, C. L. (1994). *Anticancer drugs from animals, plants, and microorganisms*. Wiley, New York.

Phillips, P., Denman, T., and Barker, S. (1997). Silent but deadly (radon gas). *Chem. Brit.*, **33**, 35.

Pratt, W. B., Ruddon, R. W., Ensminger, W. D., and Maybaum, J. (1994). *The anticancer drugs*, 2nd edn. Oxford University Press, New York.

Quigley, G. J., Wang, A. H.-J., Ughetto, G., van der Marel, G., van Boom, J. H., and Rich, A. (1980). Molecular structure of a daunomycin–DNA complex. *Proc. Natl. Acad. Sci. USA*, **77**, 7204.

Rose, J. C. *et al.*, (2003). An open randomised trial — comparison of the aromatase inhibitors letrozole and anastrozole. *Eur. J. Cancer*, **39**, 2318.

Rosenberg, B., Van Camp, L., Grimley, E. B., and Thompson, A. J. (1967). Inhibition of growth or cell division in *E. coli* by different ionic species of platinum (IV) complexes. *J. Biol. Chem.*, **242**, 1347.

Rosenberg, B., Van Camp, L., and Krigas, T. (1965). Inhibition of cell division in *E. coli* by electrolysis products from a platinum electrode. *Nature*, **205**, 698.

Russell, N. H. (1997). The biology of acute leukaemia. *Lancet*, **i**, 118.

Scott, J. D. and Williams, R. M. (2002). Ecteinascidin 743 and other tetrahydroquinoline antibiotics. *Chem. Rev.*, **102**, 1669.

Searle, C. E. (1986). Chemical carcinogens and cancer prevention. *Chem. Brit.*, **22**, 211.

Smyth, T. P., O'Donnell, M. E., O'Connor, M. J., and St. Ledger, J. O. (2000). Beta-lactamase-dependent prodrugs. *Tetrahedron*, **56**, 5699.

Sternson, S. M., Wong, J. C., Grozinger, C. M., and Schreiber, S. L. (2001). Synthesis of a structure library based on the histone deacetylase inhibitors trichostatin and trapoxin. *Org. Lett.*, **3**, 4239.

Taatjes, D. J., Gaudiano, G., Resing, K., and Koch, T. H. (1997). Redox pathway leading to the alkylation of DNA by adriamycin and daunomycin. *J. Med. Chem.*, **40**, 1276.

Takahara, P. M., Frederick, C. A., and Lipard, S. J. 1996. Crystal structure of the anticancer drug cisplatin bound to duplex DNA. *J. Amer. Chem. Soc.*, **118**, 12309.

Thomas, C. J., Rahier, N. J., and Hecht. S. M. (2004). Camptothecin — current perspectives. *Bioorg. Med. Chem.*, **12**, 1585.

(Various authors) (1996). **275**, What you need to know about cancer. *Sci. Amer.*, Sept.

(Various authors) (2002). G-protein-coupled receptors.*Chem. Biochem.*, **3**, 915–1030.

Vassilev, L. T. *et al.*, (2004). *In vivo* activation of the p53 pathway by small-molecule antagonists of MDM2 (the nutlins). *Science*, **303**, 844.

Vile, R. G. (1996). Gene therapy for cancer – hope or hype. *Chem. Ind.*, 285.

Waldmann, H., Alonso-Diaz, D., and Hinterding, K. (1998). Cell signalling pathways. *Angew. Chem. Int. Edn.*, **37**, 688.

Wall, M. E. and Wani, M. C. (1995). Camptothecin and taxol – discovery to clinic. *Cancer Res.*, **55**, 753.

Walter, S. and Buchner, J. (2002). Molecular chaperones and heat shock proteins. *Angew. Chem. Int. Edn.*, **41**, 1098.

Wang, J. C. (1985). DNA topoisomerases. *Ann. Rev. Biochem.*, **54**, 665.

Watson, W. P., Bleasdale, C., and Golding, B. T. (1996). Chemicals and cancer — estimating the risk. *Chem. Brit.*, **32**, 661

Webb, A., Cunningham, D., Cotter, F., Clarke, P. A., di Stefano, F., Ross, P., Corbo, M., and Dziewanowska, Z. (1997). *BCL-2* antisense therapy in patients with non-Hodgkin lymphoma. *Lancet*, **349**, 1137.

Weinstein-Oppenheimer, C. R., Blalock, W. L., Steelman, L. S., Chang, F., and McCubrey, J. A. (2000). The *raf* signal transduction cascade as a target for cancer chemotherapy. *Pharmacol. Therapeutics*, **88**, 229.

Wender, P. A. *et al.*, (2002). Practical synthesis of a novel and highly potent analogue of bryostatin. *J. Amer. Chem. Soc.*, **124**, 13648.

Whitmore, W. F. (1994). Localised prostate cancer — management and detection issues. *Lancet*, **i**, 1263.

Wissner, A., Floyd, M. B., Rabindran, S. K., Nilakantan, R., Greenberger, L. M., Shen, R., Wang, Y.-F., and Tsou, H.-R. (2002). Development of Iressa and Tarceva. *Bioorg, Med. Chem. Lett.*, **12**, 2893.

Zaiac, M. (2002). Taking aim at cancer (development of Glivec). *Chem. Brit.*, **38**, 44.

Zubrod, C. G. (1984). Origins of development of chemotherapy research at the NCI. *Cancer Treatment Rept.* **68**, 9.

Subject Index